CARPENTER'S CALCULATIONS MANUAL

Roger Tarbox

R. Dodge Woodson

McGRAW-HILL

New York   Chicago   San Francisco   Lisbon   London
Madrid   Mexico City   Milan   New Delhi   San Juan
Seoul   Singapore   Sydney   Toronto

The *McGraw·Hill* Companies

Cataloging-in-Publication Data is on file with the Library of Congress.

1 2 3 4 5 6 7 8 9 0   DOC/DOC   0 1 0 9 8 7 6 5 4

ISBN 0-07-143799-1

The sponsoring editor for this book was Larry S. Hager and the production supervisor was Pamela A. Pelton. It was set in Weiss by Lone Wolf Enterprises, Ltd. The art director for the cover was Anthony Landi.

Printed and bound by RR Donnelley.

McGraw-Hill books are available at special quantity discounts to use as premiums and sales promotions, or for use in corporate training programs. For more information, please write to the Director of Special Sales, McGraw-Hill Professional, Two Penn Plaza, New York, NY 10121-2298. Or contact your local bookstore.

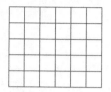

DEDICATION

I dedicate this book to Lynn. She has worked with me for many years in various functions. From dawn to dusk, she has helped me to meet deadlines and to produce quality work. Lynn's help has been instrumental in my success.

—Roger Tarbox

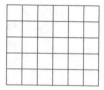

CONTENTS

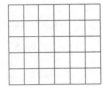

ABOUT THE AUTHORS

ROGER TARBOX

Roger Tarbox is a master carpenter with 30 years of experience. He is a native of Maine who has worked in Colorado, New Hampshire, and Maine. Mr. Tarbox has built countless homes, garages, additions, and similar structures. In addition to his work as a carpenter and contractor, he has intimate knowledge of running a sawmill and creating perfect wood products. Mr. Tarbox resides in mid-coast Maine with his wife, Lynn, and enjoys woodworking as a hobby.

R. DODGE WOODSON

R. Dodge Woodson has owned and operated his own building, remodeling, and plumbing companies since 1979. His licenses have included those of a Class A builder, a master plumber, and a master gasfitter. He has built as many as 60 homes per year and remodeled countless homes. The author of dozens of books, Mr. Woodson has a reputation for delivering technical data in a reader-friendly way. R. Dodge Woodson is the president of The Masters Group, Inc., a construction, remodeling, and plumbing firm in Maine.

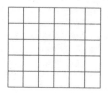

INTRODUCTION

Very few carpenters enjoy learning or doing math. If you are one of these carpenters, you are not alone. Have you ever been on a job when you needed to do math that made you think that you should have been a structural engineer if you had wanted to fiddle around with formulas? Well, the math required for building projects can be intimidating. However, there are effective shortcuts for much of it, and you will find those shortcuts in this book.

Whether you are working with windows, roofing, or wall framing, you will discover helpful, timesaving tips here. If you are new to the trade, you will appreciate the decades of experience that Roger Tarbox and R. Dodge Woodson pack into these pages. Both men have more than 20 years of experience in the building trades.

This is not a textbook for mathematics. Instead, it is a user-friendly, problem-solving guide to using math in the building trade. Much of the math is done for you in the form of conversion tables. These tables make it as simple as scanning a few columns to find solutions for your mathematical problems.

If you are looking for a no-nonsense approach to practical math applications on a job site, this is the field manual that you need. You won't find fancy formulas and long paragraphs of step-by-step instructions. The delivery of information is done quickly and effectively, so that you can work faster and smarter on a job. Take a moment to thumb through the pages. Check out the Trade Tips and the Fast Facts. Run through the conversion tables and rafter spans. It won't take long for you to see how much this material can improve your job performance.

SITE WORK

Carpenters may not think of themselves as people who need to deal with site work. Unless the carpenter is a home builder or remodeler, site work may not be much of a daily activity. Even so, there are certainly times when carpenters are involved with site work. Many of them work with this type of situation while not realizing that they are engaged in a form of site work.

What do you think of when you contemplate site work? Most carpenters probably think about cutting trees, pulling stumps, digging trenches and foundations, and so forth. All of this is site work, but the list does not cover all the aspects of it. Have you ever built a garage from the ground up? Did you deal with ground preparation for either a monolithic pour or a slab on grade? If you have done this type of work, you have done site work. How many times have you built storage sheds that rest on blocks positioned on the ground? In essence, this is also a form of site work. Admittedly, pure carpenters don't get involved with much site work on a regular basis.

Remodeling contractors and building contractors do deal with site work. Builders have to incorporate site work into all of their jobs. Remodelers who are building room additions, porches, and similar appurtenances to existing buildings also deal with site work. So, how much do you want to know about it? Even if you are not directly responsible for decisions on site improvement, it is helpful to understand the process. With this said, let's look at some helpful facts and tips that will aid you in your experiences with site conditions and site work.

SOIL PROPERTIES

Understanding soil properties will make it easier for you to establish the needs for a foundation. A little digging or test boring will provide you with samples of soil to evaluate.

TABLE 1.1 ■ Soil properties

Soil Type	Drainage Rating	Frost Heave Potential	Expansion Potential
Bedrock	Poor	Low	Low
Well-graded gravels	Good	Low	Low
Poorly graded gravels	Good	Low	Low
Well-graded sand	Good	Low	Low
Poorly graded sand	Good	Low	Low
Silty gravel	Good	Moderate	Low
Silty sand	Good	Moderate	Low
Clayey gravels	Moderate	Moderate	Low
Clayey sands	Moderate	Moderate	Low

SWELLING

The swelling and shrinkage of soil will affect a foundation. You can estimate the expected percentage of swelling and shrinkage once you know the type of soil you will be building on.

CAPACITY

What is the bearing capacity of gravel? How does this compare to clay? You should consult an engineer for exact data on the support capability for various types of soils and materials.

TRENCH SPECIFICATIONS

Whether you are digging a trench for a water service pipe or an electrical conduit, there are certain trench specifications that you should strive to meet. If a trench is going to be deep, it should be shored to prevent the trench from collapsing.

TABLE 1.2 ■ Swelling and shrinkage of soils

Type of Soil	Percent of Swell	Percent of Shrinkage
Sand	14–16	12–14
Gravel	14–16	12–14
Loam	20	17
Common earth	25	20
Dense clay	33	25
Solid rock	50–75	0

TABLE 1.3 ■ General bearing capacities of foundation soils

Material	Support Capability
Hard rock	80 tons
Loose rock	20 tons
Hardpan	10 tons
Gravel	6 tons
Coarse dry sand	3 tons
Hard clay	4 tons
Fine dry sand	3 tons
Mixed sand and clay	2 tons
Wet sand	2 tons
Firm clay	2 tons
Soft clay	1 ton

TABLE 1.4 ■ Safe loads by soil type

Tons/sq. ft. of Footing	Type of Soil
1	Soft Clay
	Sandy loam
	Firm clay/sand
	Loose fine sand
2	Hard clay
	Compact fine sand
3	Sand/gravel
	Loose coarse sand
4	Compact coarse sand
	Loose gravel
6	Gravel
	Compact sand/gravel
8	Soft rock
10	Very compact
	Gravel and sand
15	Hard pan
	Hard shale
	Sandstone
25	Medium hard rock
40	Sound hard rock
100	Bedrock
	Granite
	Gneiss

TABLE 1.5 ■ Suggested trench depths and widths

Trench Depth	Trench Width
1 foot	16 inches
2 feet	17 inches
3 feet	18 inches
4 feet	20 inches
5 feet	22 inches
6 feet*	24 inches

*Note: Trenches with depths of more than five feet should be shored up or fitted with a sheath for personal safety.

TABLE 1.6 ■ Estimating cubic yards per inch for excavation

Depth	Cubic Yards per Square Foot
2 inches	0.0006
4 inches	0.012
6 inches	0.018
8 inches	0.025
10 inches	0.031
12 inches	0.037
18 inches	0.056
24 inches	0.074
30 inches	0.111
36 inches	0.130
48 inches	0.148
52 inches	0.167
60 inches	0.185
66 inches	0.204
72 inches	0.222
78 inches	0.241
84 inches	0.259
90 inches	0.278
96 inches	0.298
102 inches	0.314

☑ *fast***facts**

When digging a trench near a footing, the trench should never be dug next to the foundation. Instead, the trench should be placed on a 45-degree angle from the foundation. This is done to prevent the chance of weakening the foundation wall or footing.

TABLE 1.7 ■ Estimating cubic yards per inch for trench excavation

Depth in Inches	Trench Width in Inches		
	12	18	24
6	1.9	2.8	3.7
12	3.7	5.6	7.4
18	5.6	8.3	11.1
24	7.4	11.1	14.8
30	9.3	13.8	18.5
36	11.1	16.6	22.2
42	13.0	19.4	25.9
48	14.8	22.2	29.6
54	16.7	25.0	33.3
60	18.6	27.8	37.0

All amounts given are based a trench length of 100 liner feet

ESTIMATING EXCAVATION VOLUME

Estimating the excavation volume for a foundation or trench is an important element in judging a fair price for a job. You can get a fairly quick estimate by using the tables shown as Table 1.6 and Table 1.7.

TREES

Whether you are a builder or remodeler, there will be times when you have to choose types of trees that are suitable for the site where you are working. If you are in charge of choosing the placement of trees, you must make sure that they are not planted in a way that they will crowd each other, or structures, as they grow. The expected height of a tree can also affect its suitability for a job. Understanding tree characteristics will help you when evaluating which trees should be eliminated from, or added to, a site.

TABLE 1.8 ■ Tree spacing

Name of Tree	Recommended Spacing (In Feet)
Douglas Fir	12
White (paper) Birch	15
Quaking Aspen	7
White Spruce	10
Red Cedar	7
Eastern White Pine	12
Pin Oak	30
Sea Grape	4
American Holly	8
Lombardy Poplar	4
Oriental Arborvitae	3
Sugar Maple	40
Red Maple	25

TABLE 1.9 ■ Tree characteristics

Type of Tree	Name of Tree	Tree Characteristics
Evergreen	Douglas Fir	Fast grower
Deciduous	White (paper) Birch	Very hardy
Deciduous	Quaking Aspen	Provides excellent visual screen
Evergreen	White Spruce	Structure breaks wind well
Evergreen	Red Cedar	Can grow well in dry soil
Evergreen	Eastern White Pine	Grows very fast
Deciduous	Pin Oak	Keeps leaves in winter
Evergreen	Sea Grape	Extremely decorative
Evergreen	American Holly	Beautiful leaves and berries
Deciduous	Lombardy Poplar	Grows fast and tall
Evergreen	Oriental Arborvitae	Grows fast
Deciduous	Sugar Maple	Beautiful tall foliage
Deciduous	Red Maple	Grows fast with great fall foliage

> ▶ *trade* **tip**

When evaluating site work, always make sure that there will be adequate drainage away from any structure. This may require hauling fill or topsoil to the site, and this can bust your budget if you don't anticipate it.

TABLE 1.10 ▪ Tree specifications

Name of Tree	Expected Height (In Feet)	Expected Width (In Feet)
Douglas Fir	60	25
White (paper) Birch	45	20
Quaking Aspen	35	5
White Spruce	45	20
Red Cedar	50	10
Eastern White Pine	70	40
Pin Oak	80	50
Sea Grape	20	8
American Holly	20	8
Lombardy Poplar	40	6
Oriental Arbovitae	16	6
Sugar Maple	80	50
Red Maple	40	30

Site work is not much of a consideration for an average carpenter, but it is a topic worth learning a bit about. In reality, most site work is determined by engineers and completed by specialized contractors. Now, let's move to the next chapter and see what needs to be known about footings.

FOOTINGS

Footings are a critical element in the proper construction of a structure. Whether you are building a shed, a garage, or a home, the footing is what will support the foundation. Granted, not all small buildings require footings. Many small storage sheds, cabins, and camps are built on a foundation of blocks that sit on top of the ground. However, most significant structures do utilize some type of footing. The most common type of footing is one that is poured in a trench to support all elements of a foundation wall.

Other types of footings are used for pier foundations, masonry chimneys, support columns, and so forth. The procedure for calculating a footing pad for a support column is different than the one used for a trench footing. Regardless of the type of footing being installed, the installation is a key element of successful building.

A trench footing can be stepped up or down to compensate for sloping land while minimizing cost. In any case, the bottom of the footing must be resting on either undisturbed solid ground or material. In some cases, the footing base might be located on compacted soil or some other suitable material. The depth of the footing will vary from region to region. This is due to the depth that frost enters the ground during winter conditions. Just as the footing is a key factor in building stability, knowing the proper depth to install the footing is equally important. A footing that is not installed at a suitable depth can be heaved by frost and cause all sorts of trouble with a building.

Most trench footings will have a concrete depth equal to the width of the foundation wall that is to be supported. The width of a trench footing will typically be twice the width of the wall being supported. A section of footing that extends on either side of a foundation wall is known as the projection of the footing.

Reinforcement bars may be installed in footings. When rebar is used, the typical size of the bar will be a half-inch diameter. It is not uncommon for

▶ *trade* **tip**

When planning a footing for a support column or chimney, the footing should be rated to carry the same weight, per square foot, of a typical footing surface.

rebar to be placed in a footing. Placement may run in crossing directions, creating squares, where the bars are tied together with wire at their connection points.

General procedure calls for a bed of crushed stone on the outside edges of footings. Slotted drainpipe is installed on the gravel bed and piped to a discharge location. This is done to keep surface water from coming into contact with the footing and building up against the foundation wall. A layer of crushed stone is installed over the drainpipe to minimize potential clogging of the pipe by soil.

FOOTING DEPTH

What is the minimum required footing depth? Local code authorities establish minimum footing depths. You should check with your local code enforcement officer to determine the minimum footing depth in your region.

TABLE 2.1 ■ Sample minimum footing depths

Extreme Winter Temperature	Minimum Footing Depth
Above freezing	1' 6"
+20 degrees F.	2' 6"
+10 degrees F.	3' 0"
0 degrees F.	3' 6"
−10 degrees F.	4' 0"
−20 degrees F.	4' 6"

▶ *trade* **tip**

Foundation footings should be installed at least 1 foot below the local frost line level. This can result in a wide variation of depth, depending upon the geographical area in which you are working. Some footings, such as in Maine, must be 4 feet deep, while footings in warmer climates might be only 18 inches deep. Consult your local code book or code officer for an exact depth to install footings in your area.

☑ *fast***facts**

Load divided by soil-bearing capacity will reveal the required number of square feet, or area, required of a foundation footing. For example, if you had a load of 60 tons per square foot that would be built in soil with a capacity rating of 20, you would need 3 square feet of footing. This means that your footing would be 3 feet wide.

CAPACITY AND AREA

The load bearing capacity of a base for a footing will play into the sizing of the footing area. Various types of soils and base materials offer different bearing capacities. You must be able to identify the bearing capacity to establish the proper size of a footing.

CALCULATING CONCRETE

Calculating concrete can be as simple as making a phone call to your concrete supplier. In most cases, if you call in the dimensions of the area that you need concrete for, a concrete supplier will compute the amount of concrete needed. This is the easy way. Keep in mind that it is better to have a little more concrete than expected rather than less than what you need. Most suppliers will factor in a percentage of excess to make sure that you have what you need, but it is wise to make sure that they do.

Having your concrete supplier figure the amount of concrete needed is fine if you are doing a job that is large enough to warrant a concrete truck. But, what if you are doing a few small footing pads or piers? Many workers mix and pour their own concrete for small jobs. Then you are on your own.

The three factors in concrete are cement, sand, and stone aggregate. If you have one cubic foot of cement, how much sand will you need? Do you know how much stone is needed?

Estimating the volume of concrete needed for a job is not difficult. It is very basic math. Multiply the length by the width by the thickness. As

▶ *trade* **tip**

- Place piers 8 feet on center when positioned perpendicular to floor joists.
- Place piers 12 feet on center when positioned parallel to floor joists.

TABLE 2.2 ■ Bearing capacities of foundation soils

Material	Support Capability	Material	Support Capability
Hard Rock	80 tons	Fine dry sand	3 tons
Loose rock	20 tons	Mixed sand and clay	2 tons
Hardpan	10 tons	Wet sand	2 tons
Gravel	6 tons	Firm clay	2 tons
Coarse dry sand	3 tons	Soft clay	1 ton
Hard clay	4 tons		

TABLE 2.3 ■ Load ratings for various soil types

Tons Per Square Foot of Footing	Types of Soils
One	Soft clay
One	Sandy loan
One	Firm clay
One	Clay
One	Sand
Two	Hard clay
Two	Compacted fine sand
Three	Sand
Three	Gravel
Three	Loose coarse sand
Four	Compacted coarse sand
Four	Loose gravel
Six	Gravel
Six	Compacted sand
Six	Compacted gravel
Eight	Soft rock
Ten	Very compacted gravel
Ten	Very compacted sand
Fifteen	Hard pan
Fifteen	Sandstone
Fifteen	Hard shale
Twenty-Five	Medium hard rock
Forty	Sound hard rock
One-hundred	Bedrock
One-hundred	Granite

▶ *trade* **tip**

Sand and stone for a concrete mix need to be clean and should not be larger in diameter than 1.5 inches.

an example, if you have 100 linear feet that will be 10 feet wide and 8 inches deep, you have an answer of about 666.6 cubic feet. The tricky part here is the depth. You are working in feet with the first two numbers. The last number, the depth, is in inches. You have to convert the depth to a fraction. If you want the fast answer, you know that 8 inches of depth out of 12 inches is about two-thirds. This works out to about 66.66. So, the formula would be 100 feet times 10 feet times 66.66. Doing this math gives you an answer of 666.6.

Once you have the cubic feet needed, you have to convert it to cubic yards. Concrete is always sold by the yard. All you have to do is divide your answer, 666.6 cubic feet, by 27. This will give you the cubic yards. In this example, the answer is 24.68 yards. There is some margin of error and you should add a little to make sure that you will have enough concrete to complete the job in one pour. So let's recount this.

Multiply the length by the width by the thickness of your pour and divide it by 27. This will give you a very close estimate of the cubic yards of concrete needed for a pour. While this may sound a little complicated, it is really pretty simple. And remember, on larger jobs, you can give your supplier the dimensions and let the supplier calculate the amount of concrete needed.

CONCRETE MIXES

There are three primary types of concrete mixes. At the low end is the economy mix. This is used in large areas that don't require a lot of strength. Then there is a mid-range mix. This type of mix is used when strength is needed and the concrete is not exposed to weather elements. If you need a mix that can be exposed to weather, wear, and water, the rating of the mix is strong.

☑ *fast***facts**

Concrete should cure for at least a week. Given a choice, it is best to let it cure for up to 28 days. The concrete should be kept moist during the curing phase. Avoid making the mixture soupy. This will result in weak concrete.

▶ *trade* **tip**

If you use bags of concrete mix from a building supplier, you can read the bag to see the yield of the mix. With a little experience, you should gain a sense of how many bags are needed for various situations. Buy a few extra bags and either keep them in inventory for future jobs or return them for credit. Always have more concrete available than you need.

TABLE 2.4 ■ Concrete mixes

Mix Type	Ratio	Material Needed Per Cubic Yard of Concrete
Economy	1:3:5	4.5 bags of cement, 13 cubic feet of sand, 22 cubic feet of stone
Mid-Range	1:2¾:4	5 bags of cement, 14 cubic feet of sand, 20 cubic feet of stone
Strong	1:2¼:3	6 bags of cement, 14 cubic feet of sand, 18 cubic feet of stone

TABLE 2.5 ■ Sand to volume conversions

- 1 cubic foot of sand is equal to approximately 100 pounds
- 1 cubic yard of sand is equal to approximately 2,700 pounds
- 1 ton of sand is equal to ¾ yard or 20 cubic feet of sand
- An average shovel filled with sand is equal to approximately 15 pounds

WATER

Water is often taken for granted, until you don't have it. Not all jobs have running water available. If you are going to mix concrete on a job where you don't have access to water on the site, you are going to have to haul it in. Putting water in a drum on a truck adds a lot of weight to the vehicle. Even a bucket of water produces considerable weight. And, how much water do you need?

TABLE 2.6 ■ Water Conversions

- 1 cubic foot of water is equal to 62.4 pounds
- 1 cubic foot of water is equal to 7.48 gallons
- 1 gallon of water is equal to 8.33 pounds

TABLE 2.7 ■ Rebar data

Bar Number	Diameter, In Inches	Square Inches of Bar Area	Approximate Weight of 100 Feet of Bar Material in Pounds
2	¼	0.05	17
3	⅜	0.11	38
4	½	0.20	67
5	⅝	0.31	104
6	¾	0.44	150

REINFORCEMENT BARS

Reinforcement bars are often a part of footings. The street name for these bars is rebar. These bars increase in size in increments of an eighth of an inch. Choosing a size depends on the tensile force needed for the concrete to carry. Someone other than a carpenter, in most cases, will determine this. Engineers and architects provide the plans for minimum construction requirements. If you prepare and pour your own footings, you will normally be working from an approved blueprint. Follow the plans and specifications as they are stated for the job that you are working on. To understand rebar ratings, refer to Table 2.7.

Footings are not a complicated part of the building process, but they are a serious element of a job. Don't cut corners when it comes to footings. Now, let's take a look at foundation factors in the next chapter.

▶ *trade* **tip**

Most jurisdictions require an inspection of the ground preparation for footings prior to the pouring of concrete. Don't forget this fact. The last thing you want is to pour a footing only to have it be required to be removed due to a lack of inspection. When the footing base is ready for concrete, get the inspections required before pouring concrete. If you are building a home with federal financing, you may need an inspection by your local code officer and by a VA (Veteran's Administration) or FHA (Federal Housing Authority) inspector, as well. Failure to obtain the multiple inspections can put you in deep trouble. The bottom line: Don't pour concrete before you have certificates of approval for required inspections.

chapter 3

FOUNDATIONS

Foundations are a major element in any successful building project. Any carpenter with field experience knows how difficult it can be to frame a structure on a foundation that is out of square. Rehabbers have seen plenty of failed foundations and the result that they have on the buildings they support. Carpenters rarely prep and pour their own large foundations, but knowing how to work with foundations is helpful to anyone involved in framing, remodeling, rehabbing, and general contracting.

Foundations take many forms. A simple pier foundation is not hard to figure out or install. But still, one needs to know how to compute the materials needed to create the foundation. Brick and block are common foundation materials. Masons will normally do their own material take-offs for these materials, but it helps when estimating job costs to have an understanding of what is involved in the estimating process. Contractors who are pouring full foundation walls need a lot of concrete. The quick way of figuring concrete is to call a concrete supplier. If you give the supplier needed measurements, the supplier can estimate the amount of concrete needed. However, many contractors prefer to depend upon themselves, rather than others, when it comes to estimating material needs.

ESTIMATING CONCRETE NEEDS

Who will be estimating your concrete needs? Are you going to do it yourself? Will your subcontractor take care of it for you? Do you depend on your supplier for estimating concrete? With a few estimating tables and formulas, you can do your own estimating. I like to run my own estimates and then have my supplier do an estimate for me. This allows me to compare the two estimates for accuracy. If there is a big discrepancy, I know to look for a problem in the figures.

☑ *fast***facts**

Multiply the length by the width by the thickness. For example, a project that is 50 feet long, 10 feet wide, and 8 inches deep will require 333.33 cubic feet of concrete. Because concrete is sold in yards, convert your findings by dividing the total (333.33) by 27. You will arrive at an answer of 12.35 cubic yards.

TABLE 3.1 ■ Concrete formulas

Grade	Ratio	Material needed for cu. yd.
Strong—watertight, exposed to weather and moderate wear	1:2 ¼:3	6 bags cement 14 cu. ft. sand (.52 yd³) 18 cu. ft. stone (.67 yd³)
Moderate—Strength, not exposed	1:2 ¾:4	5 bags cement 14 cu. ft. sand (.52 yd³) 20 cu. ft. stone (.74 yd³)
Economy—Massive areas, low strength	1:3:5	4½ bags cement 13 cu. ft. sand (.48 yd³) 22 cu. ft. stone (.82 yd³)

TABLE 3.2 ■ Quantities of concrete for footings and wall

Quantities of Concrete for Footings and Walls

Footing Size (")	Cubic Feet of concrete per LF	Cubic Yards of Concrete per 100 LF
6 x 12	0.50	1.9
8 x 12	0.67	2.5
8 x 16	0.89	3.3
10 x 12	0.83	3.1
10 x 16	1.11	4.1
10 x 18	1.25	4.6
12 x 12	1.00	3.7
12 x 24	2.00	7.4

Wall Thickness:

Thickness in inches	Square Feet	Thickness in inches	Square Feet
3	108	6	54
3½	93	6½	50
4	81	7	46
5	65	7½	43
5½	59	8	40

In addition to figuring concrete for foundation walls, you will likely deal with the need for concrete slabs. This requires a different approach.

TABLE 3.3 ■ Amount of coverage obtained from one cubic yard of concrete

One Cubic Yard of Concrete Will Place

Thickness	Square Feet
2 inches	162
2½ inches	130
3 inches	108
3½ inches	93
4 inches	81
4½ inches	72
5 inches	65
5½ inches	59
6 inches	54
6½ inches	50
7 inches	46
7½ inches	43
8 inches	40
8½ inches	38
9 inches	36
9½ inches	34
10 inches	32.5
10½ inches	31
11 inches	29.5
11½ inches	28
12 inches	27
15 inches	21.5
18 inches	18
24 inches	13.5

TABLE 3.4 ■ Recommended thickness for concrete slabs

Use	Thickness
Basement floors in residences	4 inches
Garage floors for residential use	4 to 5 inches
Porch floors	4 to 5 inches
Base for tile flooring	2½ inches
Driveways	6 to 8 inches
Sidewalks	4 to 6 inches

TABLE 3.5 ■ Estimating cubic yards of concrete needed for slabs, walks, and drives

Slab Thickness (Inches)	Slab Area (Square Feet)				
	10	50	100	300	500
2	0.1	0.3	0.6	1.9	3.1
3	0.1	0.5	0.9	2.8	4.7
4	0.1	0.6	1.2	3.7	6.2
5	0.2	0.7	1.5	4.7	7.2
6	0.2	0.9	1.9	5.6	9.3

Expansion joints are often used on walkways and should be installed in slabs that are longer than 300 feet. Reinforcements should not extend over the expansion joints.

CONCRETE MIXTURES

Concrete mixtures vary from use to use. For example, foundation walls and footings will utilize one type of mixture while a retaining wall might have a different requirement. Engineers and architects typically specify the type of mixture to be used on a large job. Smaller jobs are not always specified. This puts the burden on the contractor, installer, or supplier. Check your local code requirements and refer to manufacturers' specifications to determine proper mixtures.

TABLE 3.6 ■ Approximate weights of concrete components

Component	Volume	Approximate Weight
Cement	1 cubic foot	94 pounds
Fine, Dry Aggregate	1 cubic foot	105 pounds
Coarse Aggregate	1 cubic foot	100 pounds

TABLE 3.7 ■ Total amounts of air content percent for concrete mixes

	Exposure Levels		
Nom. Max.Aggregate Size	Severe	Moderate	Mild
⅜ inch	7.5%	6%	4.5%
½ inch	7%	5.5%	4%
¾ inch	6%	5%	3.5%
1 inch	6%	4.5%	3%
1½ inch	5.5%	4.5%	2.5%
2 inch	5%	4%	2%
3 inch	4.5%	3.5%	1.5%

Air entrained concrete has more workability than concrete without any entrainment according to one study performed by the Portland Cement Association.

REINFORCING CONCRETE

Reinforcing concrete is often a required practice. Wire mesh is almost always used in any type of slab. Steel bars, typically called rebar, is used in slabs, walls, and footings. The wire and rebar act to hold concrete together and to avoid cracking.

TABLE 3.8 ■ Welded wire mesh sizes

Wire size number		Nominal diameter, in.	Nominal weight, lb/ft	Area per width (in.²/ft) for various spacings (in)						
Plain	Deformed			2	3	4	6	8	12	16
W45	D45	0.757	1.53	2.70	1.80	1.35	0.90	0.68	0.45	0.34
W31	D31	0.628	1.05	1.86	1.24	0.93	0.62	0.47	0.31	0.23
W20	D20	0.505	0.680	1.2	0.80	0.60	0.40	0.30	0.20	0.15
W18	D18	0.479	0.612	1.1	0.72	0.54	0.36	0.27	0.18	0.14
W16	D16	0.451	0.544	0.96	0.64	0.48	0.32	0.24	0.16	0.12
W14	D14	0.422	0.476	0.84	0.56	0.42	0.28	0.21	0.14	0.11
W12	D12	0.391	0.408	0.72	0.48	0.36	0.24	0.18	0.12	0.09
W11	D11	0.374	0.374	0.66	0.44	0.33	0.22	0.17	0.11	0.08
W10.5		0.366	0.357	0.63	0.42	0.32	0.21	0.16	0.11	0.08
W10	D10	0.357	0.340	0.60	0.40	0.30	0.20	0.15	0.10	0.08
W9.5		0.348	0.323	0.57	0.38	0.29	0.19	0.14	0.095	0.07

(continued on next page)

TABLE 3.8 ■ Welded wire mesh sizes *(continued)*

| Wire size number | | Nominal diameter, in. | Nominal weight, lb/ft | Area per width (in.²/ft) for various spacings (in) | | | | | | |
Plain	Deformed			2	3	4	6	8	12	16
W9	D9	0.338	0.306	0.54	0.36	0.27	0.18	0.14	0.090	0.07
W8.5		0.329	0.289	0.51	0.34	0.26	0.17	0.13	0.085	0.06
W8	D8	0.319	0.272	0.48	0.32	0.24	0.16	0.12	0.080	0.06
W7.5		0.309	0.255	0.45	0.30	0.23	0.15	0.11	0.075	0.06
W7	D7	0.299	0.238	0.42	0.28	0.21	0.14	0.11	0.070	0.05
W6.5		0.288	0.221	0.39	0.26	0.20	0.13	0.097	0.065	0.05
W6	D6	0.276	0.204	0.36	0.24	0.18	0.12	0.090	0.060	0.05
W5.5		0.265	0.187	0.33	0.22	0.17	0.11	0.082	0.055	0.04
W5	D5	0.252	0.170	0.30	0.20	0.15	0.10	0.075	0.050	0.04
W4.5		0.239	0.153	0.27	0.18	0.14	0.090	0.067	0.045	0.03
W4	D4	0.226	0.136	0.24	0.16	0.12	0.080	0.060	0.040	0.03
W3.5		0.211	0.119	0.21	0.14	0.11	0.070	0.052	0.035	0.03
W3		0.195	0.102	0.18	0.12	0.090	0.060	0.045	0.030	0.02
W2.9		0.192	0.099	0.17	0.12	0.087	0.058	0.043	0.029	0.02
W2.5		0.178	0.085	0.15	0.10	0.075	0.050	0.037	0.025	0.02
W2.1		0.162	0.070	0.13	0.84	0.063	0.042	0.031	0.021	0.02
W2		0.160	0.068	0.12	0.080	0.060	0.040	0.030	0.020	0.02
W1.5		0.138	0.051	0.090	0.060	0.045	0.030	0.022	0.015	0.01
W1.4		0.134	0.048	0.084	0.056	0.042	0.028	0.021	0.014	0.01

(By permission, Concrete Reinforcing Steel Institute, Schramsburg, Illinois)

TABLE 3.9 ■ Types of welded wire fabric

Style designation (W = Plain, D = Deformed)	Steel area (in ²/ft)		Approximate weight (lb per 100 sq ft)
	Longitudinal	Transverse	
4 x 4-W1.4 x W1.4	0.042	0.042	31
4 x 4-W2.0 x W2.0	0.060	0.060	43
4 x 4-W2.9 x W2.9	0.087	0.087	62
4 x 4-W/D4 x W/D4	0.120	0.120	86
6 x 6-W1.4 x W1.4	0.028	0.028	21
6 x 6-W2.0 x W2.0	0.040	0.040	29
6 x 6-W2.9 x W2.9	0.058	0.058	42
6 x 6-W/D4 x W/D4	0.080	0.080	58
6 x 6-W/D4.7 x W/D4.7	0.094	0.094	68
6 x 6-W/D7.4 x W/D7.4	0.148	0.148	107
6 x 6-W/D7.5 x W/D7.5	0.150	0.150	109
6 x 6-W/D7.8 x W/D7.8	0.156	0.156	113
6 x 6-W/D8 x W/D8	0.160	0.160	116
6 x 6-W/D8.1 x W/D8.1	0.162	0.162	118
6 x 6-W/D8.3 x W/D8.3	0.166	0.166	120
12 x 12-W/D8.3 x W/D8.3	0.083	0.083	63
12 x 12-W/D8.8 x W/D8.8	0.088	0.088	67
12 x 12-W/D9.1 x W/D9.1	0.091	0.091	69
12 x 12-W/D9.4 x W/D9.4	0.094	0.094	71
12 x 12-W/D16 x W/D16	0.160	0.160	121
12 x 12-W/D16.6 x W/D16.6	0.166	0.166	126

*Many styles may be obtained in rolls.

TABLE 3.10 ■ Rebar chart

Bar Size Designation	Weight Pounds Per Foot	Nominal Dimensions-Round Sections		
		Diameter Inches	Cross-Sectional Area-Sq. Inches	Perimeter Inches
#3	.376	.375	.11	1.178
#4	.668	.500	.20	1.571
#5	1.043	.625	.31	1.963
#6	1.502	.750	.44	2.356
#7	2.044	.875	.60	2.749
#8	2.670	1.000	.79	3.142
#9	3.400	1.128	1.00	3.544
#10	4.303	1.270	1.27	3.990
#11	5.313	1.410	1.56	4.430
#14	7.650	1.693	2.25	5.320
#18	13.600	2.257	4.00	7.090

TABLE 3.11 ▪ Cure times

Cure time is a function of time, temperature, and type of cement used in the concrete mix.

The following cure times take these three factors into account.

At 50 degrees F (10 degrees C)-Measured in "days" required

Percentage design strength required	Type cement used in the mix		
	Type I	Type II	Type III
50%	6	9	3
65%	11	14	5
85%	21	28	16
95%	29	35	26

At 70 degrees F (21 degrees C) - Measured in "days" required

Percentage design strength required	Type cement used in the mix		
	Type I	Type II	Type III
50%	6	9	3
65%	11	14	5
85%	21	28	16
95%	29	35	26

CURING CONCRETE

The curing of concrete depends on various factors. Essentially, you have to determine the number of dry days needed to cure concrete. This process requires you to know the type of cement used in the concrete mix. Air temperature is also a factor when establishing a curing time.

When allowing concrete to cure, it is important that you not disturb the concrete form during the curing process. Movement of the form can weaken the concrete. Avoid using walkway staging during the curing process.

TABLE 3.12 ▪ Recommended slumps for various types of concrete construction

Unreinforced footings and substructure walls	3
Reinforced foundation walls and footings	3
Slabs	3
Pavements	3
Beams and reinforced walls	4
Columns (building)	4

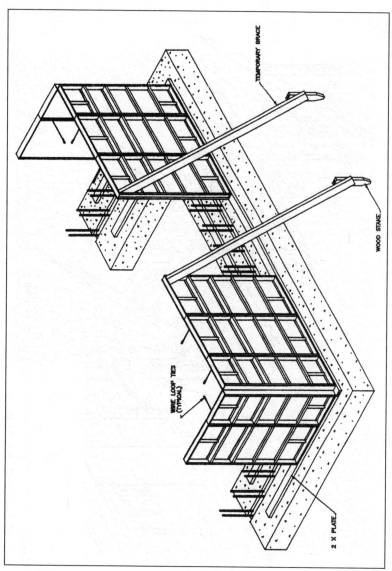

FIGURE 3.1 ■ Typical concrete wall from schematic with one side in place

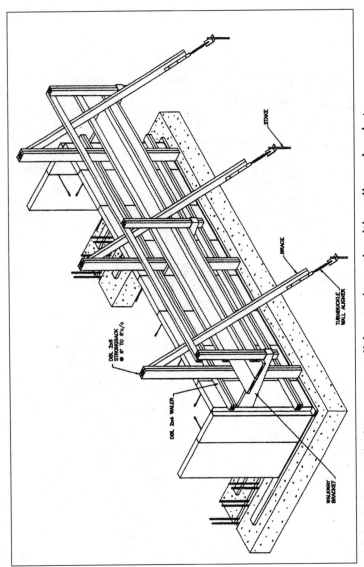

FIGURE 3.2 ■ Typical concrete wall from schematic with walkway bracket installed and one side in place

- Stretcher
- Rowlock stretcher
- Header
- Rowlock header
- Soldier

FIGURE 3.3 ■ **Types of bricks**

BRICKS AND BLOCKS

Foundation walls in many parts of the country are made with blocks and bricks. Other regions rely more on foundation walls that are poured concrete. Just as the style of wall construction materials vary, so does the use of basement walls. Many areas have most of their homes built on crawlspace foundations. Other regions feature homes with full basements. We have talked about concrete walls, so let's explore the characteristics of block and brick walls.

Bricks come in different sizes and styles. Customers, architects, or masons typically choose the type and style of brick to be used.

There are different types of brick walls that can be built. They can include:

- Cavity wall
- Brick veneer
- Single wythe
- Composite brick

Some other types of brick walls can be seen in Table 3.14.

How do you calculate the needs for building a brick or block wall? Having tables to work with can help.

TABLE 3.13 ■ **Brick sizes**

Brick type	Height (in.)	Width (in.)	Length (in.)
Standard building brick	2½	3⅞	8½
Oversize building brick	3¼	3¼	10
Norman face brick	2³⁄₁₆	3½	11½
Fire brick	2½	4½	9
Fire brick splits	1¼	4½	9

TABLE 3.14 ■ Some types of brick walls

- Flemish
- Flemish Cross
- Garden
- Common
- Common with Flemish headers
- Running
- English
- Dutch

TABLE 3.15 ■ Figuring square footage needs for 100-square-foot masonry wall

Material	Height	Units	Cement	Sand
Standard brick	4-inch wall	616	3 sacks	9 cubic feet
Standard bricks	8-inch wall	1232	7 sacks	21 cubic feet
Standard block	8-inch wall	112	1 sack	3 cubic feet

You can use tables to estimate the needs for building brick or block walls. It is a good idea to create your own tables as you track your jobs. This is the most accurate method that I know of for being on target with your estimates. Once a wall is built, count what is in it. Determine the length, width, and height, and you will have a very accurate take-off that can be broken down into a certain number of pieces per linear foot.

 *fast**facts***

To create 8 square feet of block wall, you will need nine 8-x-8-x-16 blocks. The formula for this equation is simple. Take the square footage of your proposed wall and multiply it by $1\frac{1}{8}$ (1.125) to determine your material needs.

TABLE 3.16 ■ Modular brick sizes

									Specified Dimensions Joint Thickness (Vertical)	
Unit		Dimensions (Inches)								
		Width-Height-Length				W-H-L				
Modular	4	$2\frac{2}{3}$	8		$3\frac{3}{8}$	$2\frac{1}{4}$	$7\frac{5}{8}$		$\frac{3}{8}$	3C = 8 inches
					$3\frac{1}{2}$	$2\frac{1}{4}$	$7\frac{1}{2}$		$\frac{1}{2}$	
Engineer Modular	4	$3\frac{1}{3}$	8		$3\frac{3}{8}$	$2\frac{3}{4}$	$7\frac{5}{8}$		$\frac{3}{8}$	5C = 16 inches
					$3\frac{1}{2}$	$2\frac{13}{16}$	$7\frac{1}{2}$			
Closure Modular	4	4	8		$3\frac{5}{8}$	$3\frac{5}{8}$	$7\frac{5}{8}$		$\frac{3}{8}$	1C = 4 inches
					$3\frac{1}{2}$	$3\frac{1}{2}$	$7\frac{1}{2}$		$\frac{1}{2}$	
Roman	4	2	12		$3\frac{5}{8}$	$1\frac{5}{8}$	$11\frac{5}{8}$			2C = 4 inches
					$3\frac{1}{2}$	$1\frac{1}{2}$	$11\frac{1}{2}$		$\frac{1}{2}$	
Norman	4	$2\frac{2}{3}$	12		$3\frac{5}{8}$	$2\frac{1}{4}$	$11\frac{5}{8}$		$\frac{3}{8}$	3C = 8 inches
					$3\frac{1}{2}$	$2\frac{1}{4}$	$11\frac{1}{2}$			
Engineer Norman	4	$3\frac{1}{2}$	12		$3\frac{5}{8}$	$2\frac{3}{4}$	$11\frac{5}{8}$		$\frac{3}{8}$	5C = 16 inches
					$3\frac{1}{2}$	$2\frac{13}{16}$	$11\frac{1}{2}$		$\frac{1}{2}$	
Utility	4	4	12		$3\frac{5}{8}$	$3\frac{5}{8}$	$11\frac{5}{8}$		$\frac{3}{8}$	1C = 4 inches
					$3\frac{1}{2}$	$3\frac{1}{2}$	$11\frac{1}{2}$		$\frac{1}{2}$	

Metric equivalents for modular brick sizes

$4 \times 2\frac{2}{3} \times 8 = 10.16cm \times 5.95cm \times 20.32cm$
$3\frac{3}{8} \times 2\frac{1}{4} \times 7\frac{5}{8} = 8.57cm \times 5.63 \times 18.1cm$
$3\frac{1}{2} \times 2\frac{1}{4} \times 7\frac{1}{2} = 8.75cm \times 5.63cm \times 18.75cm$

$4 \times 3 - \frac{1}{3} \times 8 = 10.16cm \times 8.33cm \times 20cm$
$3\frac{3}{8} \times 2\frac{3}{4} \times 7\frac{5}{8} = 8.57cm \times 5.63cm \times 18.1cm$
$3\frac{1}{2} \times 2\frac{13}{16} \times 7\frac{1}{2} = 8.75cm \times 5.474cm \times 18.75cm$

$4 \times 4 \times 8 = 10.16cm \times 10.16cm \times 20cm$
$3\frac{5}{8} \times 3\frac{5}{8} \times 7 - \frac{5}{8} = 9.2cm \times 9.2cm \times 18.1cm$
$3\frac{1}{2} \times 3\frac{1}{2} \times 7\frac{1}{2} = 8.75cm \times 8.75cm \times 18.75cm$

$4 \times 2 \times 12 = 10.16cm \times 5cm \times 30cm$
$3\frac{5}{8} \times 1\frac{5}{8} \times 11\frac{5}{8} = 9.2cm \times 4.127cm \times 29.52cm$
$3\frac{1}{2} \times 1\frac{1}{2} \times 11\frac{1}{2} = 8.75cm \times 3.75cm \times 28.75cm$

$4 \times 2\frac{2}{3} \times 12 = 10.16cm \times 5.95cm \times 30cm$
$3\frac{5}{8} \times 2\frac{1}{4} \times 11 - \frac{5}{8} = 9.2cm \times 5.63cm \times 29.52cm$
$3\frac{1}{2} \times 2\frac{1}{4} \times 11\frac{1}{2} = 8.75cm \times 5.63cm \times 28.75cm$

$4 \times 3\frac{1}{2} \times 12 = 10.16cm \times 8.75cm \times 30cm$
$3\frac{5}{8} \times 2\frac{3}{4} \times 11\frac{5}{8} = 9.2cm \times 5.63cm \times 29.52cm$
$3\frac{1}{2} \times 2 - \frac{13}{16} \times 11\frac{1}{2} = 8.75cm \times 5.47cm \times 28.75cm$

$4 \times 4 \times 12 = 10.16cm \times 10.16cm \times 30cm$
$3\frac{5}{8} \times 3\frac{5}{8} \times 11\frac{5}{8} = 9.2cm \times 9.2cm \times 29.52cm$
$3\frac{1}{2} \times 3\frac{1}{2} \times 11\frac{1}{2} = 8.75cm \times 8.75cm \times 28.75cm$

Note: 2C, 5C etc refers to number of courses and "inches" refers to height of that coursing.

TABLE 3.17 ■ Other modular brick sizes

Other modular brick sizes			
Nominal Size	Specified Dimensions (Inches)	Joint Thickness	Vertical Coursing
W H L	W H L		
4 6 8	3½ 5½ ½	½	2C = 12 inches
4 8 8	3½ 7½ 7½	½	1 C = 8 inches
6 3½ 12	5½ 2³⁄₁₆ 11½	½	5 C = 16 inches
6 4 12	5½ 3½ 11½	½	1C = 4 inches
8 4 12	7½ 3½ 11½	½	1 C = 4 inches
8 4 16	7½ 3½ 15½	½	1C = 4 inches

Metric equivalents for other modular brick sizes

4 x 6 x 8 = 10.16cm x 15cm x20cm
3½ x 5½ x 7½ =8.75cm x 13.75cm x 18.75cm

4 x 8 x 8 = 10.16cmx20cmx20cm
3½ x 7½ x 7½ = 8.75cm x 18.75cm x 18.75cm

6 x 3 1/2 x 12 = 15cm x 8.75cm x 30cm
5½ x 2³⁄₁₆ x 11½ =13.75cm x 5.47cmx28.75cm

6 x 4 x 12 = 15cm x 10.16cm x 30cm
5½ x 3½ x 11½ = 13.75cm x 8.75cm x 28.75cm

8 x 4 x 12 = 20cm x 10.16cm x 30cm
7½ x 3½ x 11½ = 18.75cm x 8.75cm x 28.75 cm

8 x 4x 16 = 20cm x 10.16 cm x 40 cm
7½ x 3½ x 15½ = 18.75 x 8.75cm x 38.75 cm

Other conversions for the previous page; ½" = 1.25cm, 12" = 30.48cm, 4" = 10.16cm, 8" = 20.32cm

Note: Specified dimensions may vary somewhat from manufacturer to manufacturer.

TABLE 3.18 ■ **Non-modular brick coursing**

	Non-modular brick coursing				
Unit	Specified Dimensions (Inches)			Joint Thickness	Vertical Coursing
	W	H	L		
Standard	3⅝	2¼	8	3/8	3C = 8 inches
	3½	2¼	8	½	
Engineer	3⅝	2¾	8	⅝	5C = 16 inches
Standard	3½	2¹³⁄₁₆	8	½	
Closure	3⅝	3⅝	8	3/8	1C = 4 inches
Standard	3½	3½	8	½	
King	3	2¾	9⅝	⅜	5C = 16 inches
	3	3⅝	9¾	⅜	
Queen	3	2¾	9¾	⅜	5C= 16 inches

Metric equivalents for non-modular brick coursing	
3⅝ x 2¼ x 8 = 9.20cm x 5.63cm x 20.32 cm	⅜″ = 9.5 mm
3½ x 2¼ x 8 = 8.75cm x 5.63cm x 20.32cm	½″ = 1.25 cm
3⅝ x 2¾ x 8 = 9.20cm x 6.98cm x 20.32cm	
3½ x 2¹³⁄₁₆ x 8 = 8.75cm x 5.47cm x 20.32cm	
3⅝ x 3⅝ x 8 = 9.20 cm x 9.20cm x 20.32cm	
3½ x 3½ x 8 = 8.75cm x 8.75cm x 20.32	
3 x 2¾ x 9⅝ = 7.5cm x 6.98cm x 24.08cm	
3 x 3⅝ x 9¾ = 7.5cm x 9.20 x 24.37 cm	
3 x 2¾ x 9¾ = 7.5 cm x 6.87cm x 24.37 cm	
3 x 2⅝ x 8⅝ = 7.5cm x 6.58cm x 20.16 cm	

Note: Specified dimensions may vary within this range from manufacturer to manufacturer.

TABLE 3.19 ■ Nominal height of brick and block walls by coursing

COURSES	REGULAR 4 2¼" bricks + 4 equal joints =					MODULAR 3 bricks + 3 joints =	CONCRETE BLOCKS	
	10" ¼" joints	10½" ⅜" joints	11" ½" joints	11½" ⅝" joints	12" ¾" joints	8"	3⅝" blocks ⅜" joints	7⅝" blocks ⅜" joints
1	2½"	2⅝"	2¾"	2⅞"	3"	2 11/16"	4"	8"
2	5"	5¼"	5½"	5¾"	6"	5 5/16"	8"	1'4"
3	7½"	7⅞"	8¼"	8⅝"	9"	8"	1'0"	2'0"
4	10"	10½"	11"	11½"	1'0"	10 11/16"	1'4"	2'8"
5	1'0½"	1'1⅛"	1'1¾"	1'2⅜"	1'3"	1'1 5/16"	1'8"	3'4"
6	1'3"	1'3¾"	1'4½"	1'5¼"	1'6"	1'4"	2'0"	4'0"
7	1'5½"	1'6⅜"	1'7¼"	1'8⅛"	1'9"	1'6 11/16"	2'4"	4'8"
8	1'8"	1'9"	1'10"	1'11"	2'0"	1'9 5/16"	2'8"	5'4"
9	1'10½"	1'11⅝"	2'0¾"	2'1⅞"	2'3"	2'0"	3'0"	6'0"
10	2'1"	2'2¼"	2'3½"	2'4¾"	2'6"	2'2 11/16"	3'4"	6'8"
11	2'3½"	2'4⅞"	2'6¼"	2'7⅝"	2'9"	2'5 5/16"	3'8"	7'4"
12	2'6"	2'7½"	2'9"	2'10½"	3'0"	2'8"	4'0"	8'0"
13	2'8½"	2'10⅛"	2'11¾"	3'1⅜"	3'3"	2'10 11/16"	4'4"	8'8"
14	2'11"	3'0¾"	3'2½"	3'4¼"	3'6"	3'1 5/16"	4'8"	9'4"
15	3'1½"	3'3⅜"	3'5¼"	3'7⅛"	3'9"	3'4"	5'0"	10'0"
16	3'4"	3'6"	3'8"	3'10"	4'0"	3'6 11/16"	5'4"	10'8"
17	3'6½"	3'8⅝"	3'10¾"	4'0⅞"	4'3"	3'9 5/16"	5'8"	11'4"
18	3'9"	3'11¼"	4'1½"	4'3¾"	4'6"	4'0"	6'0"	12'0"
19	3'11½"	4'1⅞"	4'4¼"	4'6⅝"	4'9"	4'2 11/16"	6'4"	12'8"
20	4'2"	4'4½"	4'7"	4'9½"	5'0"	4'5 5/16"	6'8"	13'4"
21	4'4½"	4'7⅛"	4'9¾"	5'0⅜"	5'3"	4'8"	7'0"	14'0"
22	4'7"	4'9¾"	5'0½"	5'3¼"	5'6"	4'10 11/16"	7'4"	14'8"
23	4'9½"	5'0⅜"	5'3¼"	5'6⅛"	5'9"	5'1 5/16"	7'8"	15'4"
24	5'0"	5'3"	5'6"	5'9"	6'0"	5'4"	8'0"	16'0"
25	5'2½"	5'5⅝"	5'8¾"	5'11⅞"	6'3"	5'6 11/16"	8'4"	16'8"

26	5' 5"	5' 8¼"	5' 11½"	6' 2¾"	6' 6"	5' 9 9/16"	8' 8"	17' 4"
27	5' 7½"	5' 10⅞"	6' 2¼"	6' 5⅝"	6' 9"	6' 0"	9' 0"	18' 0"
28	5' 10"	6' 1½"	6' 5"	6' 8½"	7' 0"	6' 2 11/16"	9' 4"	18' 8"
29	6' 0½"	6' 4⅛"	6' 7¾"	6' 11⅜"	7' 3"	6' 5 5/16"	9' 8"	19' 4"
30	6' 3"	6' 6¾"	6' 10½"	7' 2¼"	7' 6"	6' 8"	10' 0"	20' 0"
31	6' 5½"	6' 9⅜"	7' 1¼"	7' 5⅛"	7' 9"	6' 10 11/16"	10' 4"	20' 8"
32	6' 8"	7' 0"	7' 4"	7' 8"	8' 0"	7' 1 5/16"	10' 8"	21' 4"
33	6' 10½"	7' 2⅝"	7' 6¾"	7' 10⅞"	8' 3"	7' 4"	11' 0"	22' 0"
34	7' 1"	7' 5¼"	7' 9½"	8' 1¾"	8' 6"	7' 6 11/16"	11' 4"	22' 8"
35	7' 3½"	7' 7⅞"	8' 0¼"	8' 4⅝"	8' 9"	7' 9 5/16"	11' 8"	23' 4"
36	7' 6"	7' 10½"	8' 3"	8' 7½"	9' 0"	8' 0"	12' 0"	24' 0"
37	7' 8½"	8' 1⅛"	8' 5¾"	8' 10⅜"	9' 3"	8' 2 11/16"	12' 4"	24' 8"
38	7' 11"	8' 3¾"	8' 8½"	9' 1¼"	9' 6"	8' 5 5/16"	12' 8"	25' 4"
39	8' 1½"	8' 6⅜"	8' 11¼"	9' 4⅛"	9' 9"	8' 8"	13' 0"	26' 0"
40	8' 4"	8' 9"	9' 2"	9' 7"	10' 0"	8' 10 11/16"	13' 4"	26' 8"
41	8' 6½"	8' 11⅝"	9' 4¾"	9' 9⅞"	10' 3"	9' 1 5/16"	13' 8"	27' 4"
42	8' 9"	9' 2¼"	9' 7½"	10' 0¾"	10' 6"	9' 4"	14' 0"	28' 0"
43	8' 11½"	9' 4⅞"	9' 10¼"	10' 3⅝"	10' 9"	9' 6 11/16"	14' 4"	28' 8"
44	9' 2"	9' 7½"	10' 1"	10' 6½"	11' 0"	9' 9 5/16"	14' 8"	29' 4"
45	9' 4½"	9' 10⅛"	10' 3¾"	10' 9⅜"	11' 3"	10' 0"	15' 0"	30' 0"
46	9' 7"	10' 0¾"	10' 6½"	11' 0¼"	11' 6"	10' 2 11/16"	15' 4"	30' 8"
47	9' 9½"	10' 3⅜"	10' 9¼"	11' 3⅛"	11' 9"	10' 5 5/16"	15' 8"	31' 4"
48	10' 0"	10' 6"	11' 0"	11' 6"	12' 0"	10' 8"	16' 0"	32' 0"
49	10' 2½"	10' 8⅝"	11' 2¾"	11' 8⅞"	12' 3"	10' 10 11/16"	16' 4"	32' 8"
50	10' 5"	10' 11¼"	11' 5½"	11' 11¾"	12' 6"	11' 1 5/16"	16' 8"	33' 4"

TABLE 3.20 ■ Length of CMU walls by stretcher and height by coursing

Estimating Concrete Masonry

NOMINAL LENGTH OF CONCRETE MASONRY WALLS BY STRETCHERS
(Based on units 15⅝" long and half units 7⅝" long with ⅜" thick head joints)

LENGTH OF WALL	NO. OF UNITS	LENGTH OF WALL	NO. OF UNITS	LENGTH OF WALL	NO. OF UNITS	LENGTH OF WALL	NO. OF UNITS	LENGTH OF WALL	NO. OF UNITS	LENGTH OF WALL	NO. OF UNITS
0'-8"	½	20'-8"	15½	40'-8"	30½	60'-8"	45½	80'-8"	60½	100'-8"	75½
1'-4"	1	21'-4"	16	41'-4"	31	61'-4"	46	81'-4"	61	101'-4"	76
2'-0"	1½	22'-0"	16½	42'-0"	31½	62'-0"	46½	82'-0"	61½	102'-0"	76½
2'-8"	2	22'-8"	17	42'-8"	32	62'-8"	47	82'-8"	62	102'-8"	77
3'-4"	2½	23'-4"	17½	43'-4"	32½	63'-4"	47½	83'-4"	62½	103'-4"	77½
4'-0"	3	24'-0"	18	44'-0"	33	64'-0"	48	84'-0"	63	104'-0"	78
4'-8"	3½	24'-8"	18½	44'-8"	33½	64'-8"	48½	84'-8"	63½	104'-8"	78½
5'-4"	4	25'-4"	19	45'-4"	34	65'-4"	49	85'-4"	64	105'-4"	79
6'-0"	4½	26'-0"	19½	46'-0"	34½	66'-0"	49½	86'-0"	64½	106'-0"	79½
6'-8"	5	26'-8"	20	46'-8"	35	66'-8"	50	86'-8"	65	106'-8"	80
7'-4"	5½	27'-4"	20½	47'-4"	35½	67'-4"	50½	87'-4"	65½	107'-4"	80½
8'-0"	6	28'-0"	21	48'-0"	36	68'-0"	51	88'-0"	66	108'-0"	81
8'-8"	6½	28'-8"	21½	48'-8"	36½	68'-8"	51½	88'-8"	66½	108'-8"	81½
9'-4"	7	29'-4"	22	49'-4"	37	69'-4"	52	89'-4"	67	109'-4"	82
10'-0"	7½	30'-0"	22½	50'-0"	37½	70'-0"	52½	90'-0"	67½	110'-0"	82½
10'-8"	8	30'-8"	23	50'-8"	38	70'-8"	53	90'-8"	68	110'-8"	83
11'-4"	8½	31'-4"	23½	51'-4"	38½	71'-4"	53½	91'-4"	68½	111'-4"	83½
12'-0"	9	32'-0"	24	52'-0"	39	72'-0"	54	92'-0"	69	112'-0"	84
12'-8"	9½	32'-8"	24½	52'-8"	39½	72'-8"	54½	92'-8"	69½	112'-8"	84½
13'-4"	10	33'-4"	25	53'-4"	40	73'-4"	55	93'-4"	70	113'-4"	85
14'-0"	10½	34'-0"	25½	54'-0"	40½	74'-0"	55½	94'-0"	70½	114'-0"	85½

HOW TO USE THESE TABLES

The tables on this page are an aid to estimating and designing with standard concrete masonry units. The following are examples of how they can be used to advantage.

Example:
Estimate the number of units required for a wall 76' long and 12' high.

From table: 76' = 57 units
12' = 18 courses
57 × 18 = 1026 = No. masonry units required

Example:
Estimate the number of units required for a foundation 24' × 30' × 11 courses high.
2 (24 + 30) = 108' = distance for a foundation
From table: 108' = 81 units
81 × 11 = 891 = No. masonry units required.

This table can also be useful in the layout of a building on a modular basis to eliminate cutting of units. Example: If design calls for a wall 41' long it can be found from the table that making wall 41'-4", will eliminate cutting units and consequent waste. Example: If the distance between two openings has been tentatively established at 2'-9", consulting the table will show that 2'-8" dimension would eliminate cutting of units.

NOMINAL HEIGHT OF CONCRETE MASONRY WALLS BY COURSES
(Based on units 7⅝" high ⅜" thick mortar joints)

HEIGHT OF WALL	NO. OF UNITS	HEIGHT OF WALL	NO. OF UNITS	HEIGHT OF WALL	NO. OF UNITS	HEIGHT OF WALL	NO. OF UNITS
0'-8"	1	8'-8"	13	16'-8"	25	24'-8"	37
1'-4"	2	9'-4"	14	17'-4"	26	25'-4"	38
2'-0"	3	10'-0"	15	18'-0"	27	26'-0"	39
2'-8"	4	10'-8"	16	18'-8"	28	26'-8"	40
3'-4"	5	11'-4"	17	19'-4"	29	27'-4"	41
4'-0"	6	12'-0"	18	20'-0"	30	28'-0"	42
4'-8"	7	12'-8"	19	20'-8"	31	28'-8"	43
5'-4"	8	13'-4"	20	21'-4"	32	29'-4"	44
6'-0"	9	14'-0"	21	22'-0"	33	30'-0"	45
6'-8"	10	14'-8"	22	22'-8"	34	30'-8"	46
7'-4"	11	15'-4"	23	23'-4"	35	31'-4"	47
8'-0"	12	16'-0"	24	24'-0"	36	32'-0"	48

HEIGHT OF WALL	NO. OF UNITS	HEIGHT OF WALL	NO. OF UNITS	HEIGHT OF WALL	NO. OF UNITS	HEIGHT OF WALL	NO. OF UNITS	HEIGHT OF WALL	NO. OF UNITS	HEIGHT OF WALL	NO. OF UNITS
14'-8"	11	34'-8"	26	54'-8"	41	74'-8"	56	94'-8"	71	114'-8"	86
15'-4"	11½	35'-4"	26½	55'-4"	41½	75'-4"	56½	95'-4"	71½	115'-4"	86½
16'-0"	12	36'-0"	27	56'-0"	42	76'-0"	57	96'-0"	72	116'-0"	87
16'-8"	12½	36'-8"	27½	56'-8"	42½	76'-8"	57½	96'-8"	72½	116'-8"	87½
17'-4"	13	37'-4"	28	57'-4"	43	77'-4"	58	97'-4"	73	117'-4"	88
18'-0"	13½	38'-0"	28½	58'-0"	43½	78'-0"	58½	98'-0"	73½	118'-0"	88½
18'-8"	14	38'-8"	29	58'-8"	44	78'-8"	59	98'-8"	74	118'-8"	89
19'-4"	14½	39'-4"	29½	59'-4"	44½	79'-4"	59½	99'-4"	74½	119'-4"	89½
20'-0"	15	40'-0"	30	60'-0"	45	80'-0"	60	100'-0"	75	120'-0"	90

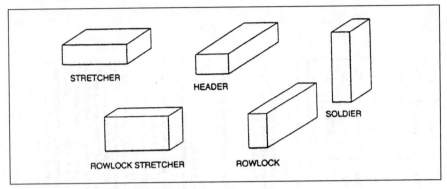

FIGURE 3.4 ■ Brick positions in a wall

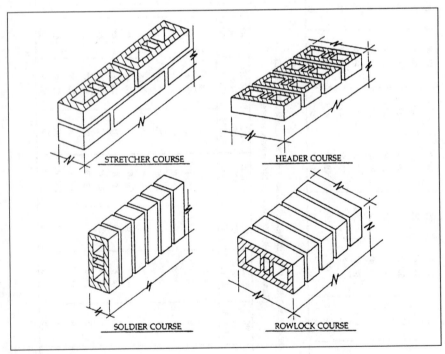

FIGURE 3.5 ■ Brick orientation

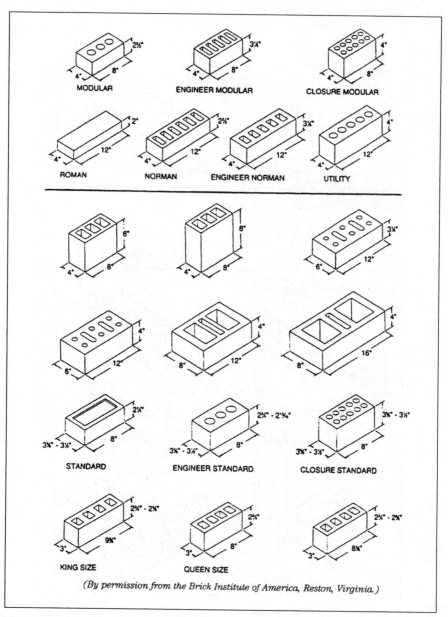

FIGURE 3.6 ▪ Modular (above) and nonmodular (below) brick sizes

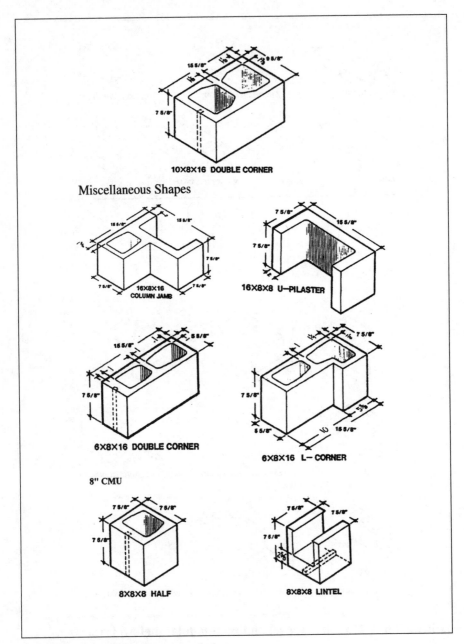

FIGURE 3.7 ■ CMU shapes and sizes

TYPES OF BLOCKS

What types of blocks will be used in your foundation walls? There are many to choose from. A mason will normally determine the material needs for a block wall, but it never hurts to understand the estimates that you are reviewing and working with. To do this, it helps to understand the types of blocks that may be used.

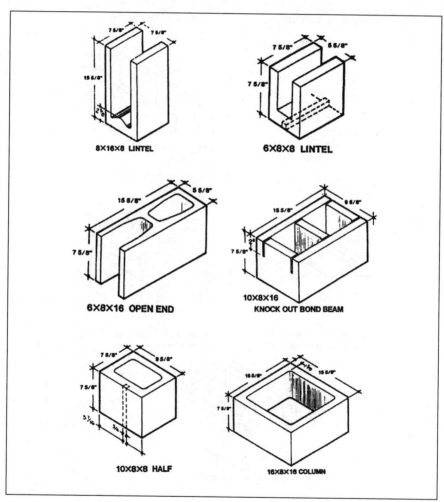

FIGURE 3.8 ■ CMU shapes and sizes

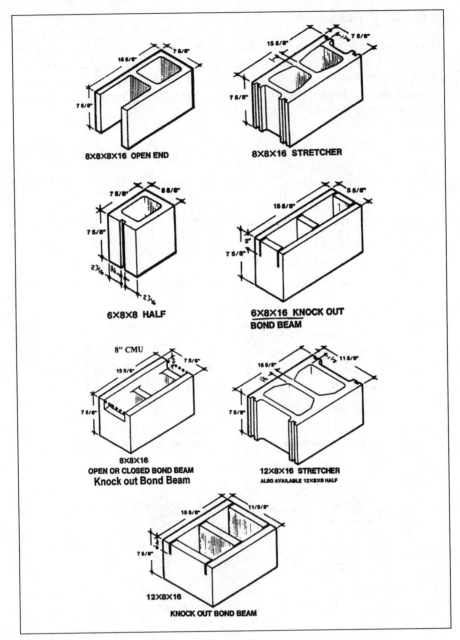

FIGURE 3.9 ■ CMU shapes and sizes

TABLE 3.21 ■ Crawlspace gross vent area requirements

	Multiply free vent area by	
Vent cover material	With soil cover	No soil cover
¼" mesh hardware cloth	1.0	10
⅛" mesh screen	1.25	12.5
16-mesh insect screen	2.0	20
Louvers + ¼" hardware cloth	2.0	20
Louvers + ⅛" mesh screen	2.25	22.5
Louvers + 16-mesh screen	3.0	30

Crawlspace foundations require ventilation. This is true of brick and block walls that are used to create crawlspace foundations. You will need to confirm that your mason will leave holes of the appropriate size in the proper locations for adequate ventilation.

MORTAR

Bricks and blocks are held together by mortar. This is no big secret. Some types of mortar joints are weather resistant, while others are not. Do you know which are and which are not?

Not weather-resistant
- Flush
- Raked
- Struck

Weather-resistant
- Concave
- V-shaped
- Weathered

FIGURE 3.10 ■ Facts about mortar joints

TABLE 3.22 ■ Mortar ratings

Mortar	Rated for
Type M	Vigorous exposure, load-bearing capability, and below-grade use
Type S	Severe exposure, load-bearing capability, and below-grade use
Type N	Light loads, mild exposure, and above-grade use
Type O	Light loads and interior use
Type K	Nonbearing use or for where compressive strength does not exceed 75 PSI

TABLE 3.23 ■ Mortar statistics

Compressive	Mortar strength (PSI)
Type M	2500
Type S	1800
Type N	750
Type O	350
Type K	75

STRESS FACTORS

Stress factors have to be taken into consideration when working with foundation walls. Compressive stress is allowable, but within limits.

Since you are probably a carpenter, you may be pleased to know that we are now ready to move to the next chapter and talk about sills and joists. This topic may be more to your liking. However, what you have learned to this point will serve you well in the field.

TABLE 3.24 ■ Allowable compressive stresses for masonry

Construction; compressive strength of unit, gross area, psi (MPa)	Allowable compressive stresses[1] gross cross-sectional area, psi (MPa)	
	Type M or S mortar	Type N mortar
Solid masonry of brick and other solid units of clay or shale; sand-lime or concrete brick:		
8000 (55.1) or greater	350 (2.4)	300 (2.1)
4500 (31.0)	225 (1.6)	200 (1.4)
2500 (17.2)	160 (1.1)	140 (0.97)
1500 (10.3)	115 (0.79)	100 (0.69)
Grouted masonry, of clay or shale; sand-lime or concrete:		
4500 (31.0) or greater	225 (1.6)	200 (1.4)
2500 (17.2)	160 (1.1)	140 (0.97)
1500 (8.3)	115 (0.79)	100 (0.69)
Solid masonry of solid concrete masonry units:		
3000 (20.7) or greater	225 (1.6)	200 (1.4)
2000 (13.8)	160 (1.1)	140 (0.97)
1200 (8.3)	115 (0.79)	100 (0.69)
Masonry of hollow load bearing units:		
2000 (13.8) or greater	140 (0.97)	120 (0.83)
1500 (10.3)	115 (0.79)	100 (0.69)
1000 (6.9)	75 (0.52)	70 (0.48)
700 (4.8)	60 (0.41)	55 (0.38)
Hollow walls (noncomposite masonry bonded) Solid units:		
2500 (17.2) or greater	160 (1.1)	140 (0.97)
1500 (10.3)	115 (0.79)	100 (0.69)
Hollow units	75 (0.52)	70 (0.48)
Stone ashlar masonry:		
Granite	720 (5.0)	640 (4.4)
Limestone or marble	450 (3.1)	400 (2.8)
Sandstone or cast stone	360 (2.5)	320 (2.2)
Rubble stone masonry Coursed, rough, or random	120 (0.83)	100 (0.69)

(continued)

TABLE 3.24 ■ Allowable compressive stresses for masonry (continued)

Net area compressive strength of units, psi (MPa)	Moduli of elasticity[1] E, psi × 10⁶ (MPa × 10³)	
	Type N mortar	Type M or S mortar
6000 (41.3) and greater	—	3.5 (24)
5000 (34.5)	2.8 (19)	3.2 (22)
4000 (27.6)	2.6 (18)	2.9 (20)
3000 (20.7)	2.3 (16)	2.5 (17)
2500 (17.2)	2.2 (16)	2.4 (17)
2000 (13.8)	1.8 (12)	2.2 (15)
1500 (10.3)	1.5 (10)	1.6 (11)

[1]Linear interpolation permitted.

(By permission from the Masonry Society, ACI, ASCE from their manual Building Code Requirements for Masonry Structures.)

TABLE 3.25 ■ Compressive strength of mortars made with various types of cement

Type of cement	Minimum compressive strength, psi				ASTM designation
	1 day	3 days	7 days	28 days	
Portland cements					C150-85
I	—	1800	2800	4000*	
IA	—	1450	2250	3200*	
II	—	1500	2500	4000*	
	—	1000†	1700†	3200*†	
IIA	—	1200	2000	3200*	
	—	800†	1350†	2560*†	
III	1800	3500	—	—	
IIIA	1450	2800	—	—	
IV	—	—	1000	2500	
V	—	1200	2200	3000	
Blended cements					C595-85
I(SM), IS, I(PM), IP	—	1800	2800	3500	
I(SM)-A, IS-A I(PM)-A, IP-A	—	1450	2250	2800	
IS(MS), IP(MS)	—	1500	2500	3500	
IS-A(MS), IP-A(MS)	—	1200	2000	2800	
S	—	—	600	1500	
SA	—	—	500	1250	
P	—	—	1500	3000	
PA	—	—	1250	2500	
Expansive cement					C845-80
E-1	—	—	2100	3500	
Masonry cements					C91-83a
N	—	—	500	900	
S	—	—	1300	2100	
M	—	—	1800	2900	

*Optional requirement.
†Applicable when the optional heat of hydration or chemical limit on the sum of C2S and C3A is specified.
Note: When low or moderate heat of hydration is specified for blended cements (ASTM C595), the strength requirements is 80% of the value shown.
(By permission from the Masonry Society, ACI, ASCE from their manual Building Code Requirements for Masonry Structures.)

chapter 4

SILLS AND JOISTS

Sills and joists are the starting point for most carpenters. This is where the wood meets the foundation. The installation of these framing members is not normally complicated, but it is necessary to size the structural elements properly. This amounts to knowing how much distance a joist can span without direct support. The allowable length of a joist varies with different types of wood. Another factor that affects the length of joists is the horizontal spacing of joists.

It is most common to find joists set 16 inches on center. However, joists are sometimes spaced at 24 inches on center. There are other options for horizontal spacing. In any event, this spacing affects the overall allowable length of a joist between supports.

Another factor that affects the developed length of a joist is the live load rating for what is to be supported by the joist. For example, a floor with a live load of 40 will require shorter joist spans than if the floor was to have a rating of 30 for the live load. Local building codes define the usage of space and the acceptable load ratings.

Dead loads are another factor when sizing joists. A dead load is determined by the weight of building materials, such as subflooring and floor coverings. When dealing with dead loads, the ratings will be based on pounds per square foot (PSF).

Sills are usually placed on a sill sealer, which is normally a lightweight material that fits under the sill and on top of the foundation wall. It is common for sills to be secured to anchor bolts that were placed in the foundation wall during its construction.

Joists are used to support floors, but they can also be used to create a ceiling when there are multiple stories in the chosen type of construction. When joists double for use in holding a ceiling and a floor, they are typically called ceiling joists.

TABLE 4.1 ■ Typical maximum floor joist spans with a 40-pound live load rating

Joist size (inches)	Spacing (inches)	Pine/fir (feet/inches)
2 × 6	12	10-6
2 × 6	16	9-8
2 × 6	24	8-4
2 × 8	12	14-4
2 × 8	16	13-0
2 × 8	24	10-4
2 × 10	12	17-4
2 × 10	16	16-2
2 × 10	24	14-6
2 × 12	12	20-0
2 × 12	16	18-8
2 × 12	24	16-10

TABLE 4.2 ■ Minimum uniformly distributed live Loads

- The first floor of a residential dwelling should be rated at 40 pounds per square foot (PSF).
- Other floors should be rated at 30 PSF.
- Stair treads should be rated at 75 PSF.
- Roofs used for sun decks should be rated at 30 PSF.
- Garages for passenger cars should be rated at 50 PSF.
- Attics accessible by stairs or ladder should be rated at 30 PSF, when the ceiling height is more than 4½ feet.
- Attics accessible by scuttle hole with a ceiling height of less than 4½ feet can be rated at 20 PSF.

TABLE 4.3 ■ Residential live loads

Area/activity	Live load, PSF
First floor	40
Second floor and habitable attics	30
Balconies, fire escapes, and stairs	100
Garages	50

TABLE 4.4 ■ Dead load weights of various flooring materials

Material	Load per PSF
Standard 2 × 4 (16" on center)	2
Standard 2 × 6 (16" on center)	2
Standard 2 × 8 (16" on center)	3
Standard 2 × 10 (16" on center)	3
Softwood, per inch of thickness	3
Hardwood, per inch of thickness	4
Plywood, per inch of thickness	3
Concrete, per inch of thickness	12
Stone, per inch of thickness	13
Carpet, per inch of thickness	0.5

▶ *trade* **tip**

To find a dead-load rate, you have to calculate all building materials and their effective weight ratios. For example, if you have a building that has 2 x 8 floor joists, set 16 inches on center, that are covered with ½" plywood and carpet, you have a dead load of 5 PSF. This solution is arrived at by taking the rating of the joists (3 PSF) plus the weight of the plywood (1.5 PSF) and the carpeting (0.5 PSF).

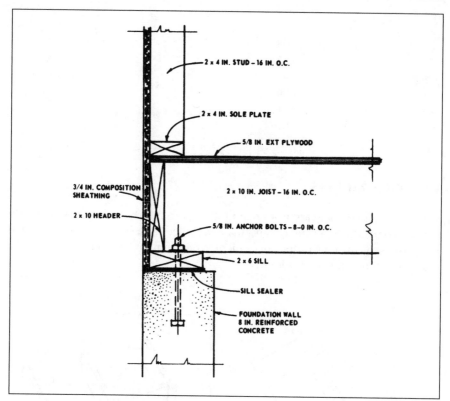

FIGURE 4.1 ▪ Framing detail

☑ *fast***facts**

When calculating loads on ceiling joists, consider what the space above the joists will be used for. The use may be as a floor for finished rooms. Or, it might be a simple attic with no storage capacity factored into the weight ratings. Then there are attics that are used for storage, and this type of use will require a calculation for the weight of that storage use.

TABLE 4.5 ▪ Typical maximum ceiling joist spans without attic storage capacity

Joist size (inches)	Spacing (inches)	Pine (feet/inches)
2 × 6	12	11-10
2 × 6	16	10-10
2 × 6	24	9-6
2 × 8	12	17-2
2 × 8	16	16-0
2 × 8	24	14-4
2 × 10	12	21-8
2 × 10	16	20-2
2 × 10	24	18-4
2 × 12	12	24-0
2 × 12	16	23-10
2 × 12	24	21-10

FRAMING PRACTICES

Framing practices can vary from region to region, but they are generally pretty similar. For example, bridging is normally installed between joists to ensure that the joists stay in their proper placement. Some builders prefer cross bridging while others prefer solid bridging.

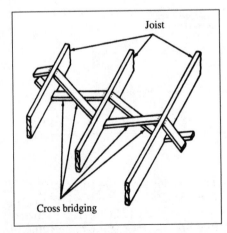

FIGURE 4.2 ▪ Cross bridging
(Courtesy U.S. government)

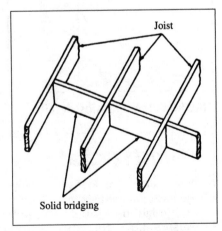

FIGURE 4.3 ▪ Solid bridging
(Courtesy U.S. government)

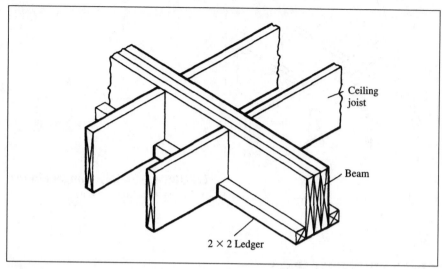

FIGURE 4.4 ▪ **The use of a ledger support with a wooden beam**
(Courtesy U.S. government)

The types of beams used can range from basic wood to engineered wood beams, to steel. All of the choices are fine, so long as they are determined based on usage and code requirements. Depending on the type of beam used, some carpenters will use ledger strips to support joists while other carpenters prefer to use metal joist hangers. I have found metal joist hangers to be the choice of most builders when they are working with beams made of dimensional lumber.

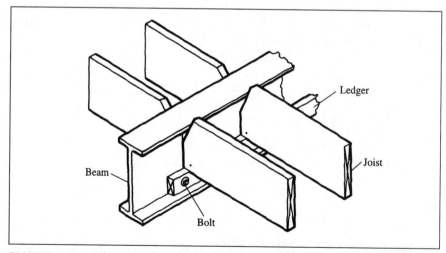

FIGURE 4.5 ▪ **The use of a ledger support with a steel beam**
(Courtesy U.S. government)

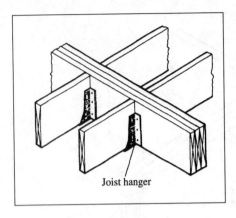

Joist hanger

FIGURE 4.6 ■ Joist hangers in use
(Courtesy U.S. government)

CALCULATING JOIST SPANS

Calculating joist spans is not difficult when you have suitable tables to work with. You won't need a lot of math ability to refer to a sizing table and determine maximum joists lengths with various types of wood. Weight, both in live loads and dead loads, is an essential part of the sizing process. Refer to your local building code to determine acceptable weight ratings.

How much weight will a girder hold? It depends, of course, on the type of girder being used. The size and type of girder to be used will normally be designed by a design professional. It is unlikely that a carpenter in the field will have to compute maximum girder size. However, if you know the weight to be supported and have a conversion table, such as what you will find in most codebooks, the process is not difficult. Refer to Tables 4.7 and 4.8 to see examples of the types of tables that you can work with. However, check your local code for the precise ratings in your region.

☑ *fast***facts**

Board Measure of Dimensional Lumber

Nominal Size	Board Feet Per Linear Foot
2 x 6	1
2 x 8	$1\frac{1}{3}$
2 x 10	$1\frac{2}{3}$
2 x 12	2

TABLE 4.6 ■ Typical live load ratings

Type of Use	Live Load (in pounds per square feet)
Bedroom	30
Residential rooms (other than bedrooms)	40
Ceiling joists without attic use	5
Ceiling joist with light storage use	20
Attic room joists	30

The type of wood you use for joists will influence the maximum length of an unsupported section of joist. Tables 4.9, 4.10, and 4.11 show examples of variances in wood types and spans at different load ratings.

WOOD SPECIES

What wood species do you normally work with? It will be necessary to determine the species of wood to be used as joists before you can calculate the allowable joist span. For example, Eastern Hemlock will allow a greater joist span than Eastern Spruce will. The tables that follow will give you a wealth of data on joist spans for different species of wood.

TABLE 4.7 ■ Wood loads on 6-inch and 8-inch girders, based on 1,200 pounds per square inch in fiber stress

The following table shows the maximum uniformly distributed load, in pounds, for the size and span of various girder conditions.

Girder Size	Span in Feet	Maximum Uniform Load of Weight
2 x 6	6	1125
2 x 6	7	960
2 x 6	8	835
6 x 6	6	3650
6 x 6	7	3110
6 x 6	8	2710
2 x 8	6	1605
2 x 8	7	1605
2 x 8	8	1505
2 x 8	9	1330
2 x 8	10	1190
2 x 8	11	1075

TABLE 4.8 ■ Wood loads on 10-inch and 12-inch girders, based on 1,200 pounds per square inch in fiber stress

The following table shows the maximum uniformly distributed load, in pounds, for the size and span of various girder conditions.

Girder Size	Span in Feet	Maximum Uniform Load of Weight
2 x 10	8	2020
2 x 10	9	2020
2 x 10	10	1910
2 x 10	11	1730
2 x 10	12	1580
2 x 10	13	1450
2 x 10	14	1335
2 x 12	11	2435
2 x 12	12	2330
2 x 12	13	2135
2 x 12	14	1975
2 x 12	15	1830
2 x 12	16	1710

TABLE 4.9 ■ Acceptable spans for Grade Number 2 construction joists

Table refers to floor joists with a live load of 40 pounds per square foot

Wood Species	Joist Size	16 inches on center	24 inches on center
Balsam Fir	2 x 6	9' 1"	7' 9"
Balsam Fir	2 x 8	12'	10' 2"
Balsam Fir	2 x 10	15' 3"	13'
Balsam Fir	2 x 12	18' 7"	15' 10"
Douglas Fir-Larch	2 x 6	9' 11"	8' 6"
Douglas Fir-Larch	2 x 8	13' 1"	11' 3"
Douglas Fir-Larch	2 x 10	16' 9"	14' 4"
Douglas Fir-Larch	2 x 12	20' 4"	17' 5"
Douglas Fir-South	2 x 6	9' 1"	7' 11"
Douglas Fir-South	2 x 8	12'	10' 6"
Douglas Fir-South	2 x 10	15' 3"	13' 4"
Douglas Fir-South	2 x 12	18' 7"	16' 3"
Eastern White Pine	2 x 6	8' 4"	6' 9"
Eastern White Pine	2 x 8	11'	8' 11"
Eastern White Pine	2 x 10	14'	11'' 4"
Eastern White Pine	2 x 12	17'	13' 10"

TABLE 4.10 ■ Acceptable spans for Grade Number 2 construction joists

Table refers to floor joists with a live load of 30 pounds per square foot

Wood Species	Joist Size	16 inches on center	24 inches on center
Balsam Fir	2 x 6	10'	8' 6"
Balsam Fir	2 x 8	13' 2"	11' 3"
Balsam Fir	2 x 10	16' 10"	14' 4"
Balsam Fir	2 x 12	20' 6"	17' 5"
Douglas Fir-Larch	2 x 6	10' 11"	9' 7"
Douglas Fir-Larch	2 x 8	14' 5"	12' 7"
Douglas Fir-Larch	2 x 10	18' 5"	16' 1"
Douglas Fir-Larch	2 x 12	22' 5"	19' 7"
Douglas Fir-South	2 x 6	10'	8' 9"
Douglas Fir-South	2 x 8	13' 2"	11' 6"
Douglas Fir-South	2 x 10	16' 10"	14' 8"
Douglas Fir-South	2 x 12	20' 6"	17' 11"
Eastern White Pine	2 x 6	9' 6"	7' 9"
Eastern White Pine	2 x 8	12' 6"	10' 2"
Eastern White Pine	2 x 10	15' 11"	13'
Eastern White Pine	2 x 12	19' 4"	15' 10"

TABLE 4.11 ■ Acceptable spans for Grade Number 2 construction joists

Table refers to floor joists with a live load of 20 pounds per square foot

Wood Species	Joist Size	16 inches on center	24 inches on center
Balsam Fir	2 x 6	11' 5"	10'
Balsam Fir	2 x 8	15' 1"	13' 2"
Balsam Fir	2 x 10	19' 3"	16' 10"
Douglas Fir-Larch	2 x 6	12' 6"	10' 11"
Douglas Fir-Larch	2 x 8	16' 6"	14' 5"
Douglas Fir-Larch	2 x 10	16' 9"	14' 4"
Douglas Fir-South	2 x 6	11' 5"	10'
Douglas Fir-South	2 x 8	15' 1"	13' 2"
Douglas Fir-South	2 x 10	19' 3"	16' 10"
Eastern White Pine	2 x 6	10' 10"	8' 10"
Eastern White Pine	2 x 8	14' 3"	11' 8"
Eastern White Pine	2 x 10	18' 3"	14' 11"

TABLE 4.12 ■ Acceptable spans for Grade Number 2 construction joists

Wood Species	Joist Size	16 inches on center	24 inches on center
Table refers to floor joists with a live load of 40 pounds per square foot			
Eastern Spruce	2 x 6	8' 4"	6' 9"
Eastern Spruce	2 x 8	11'	8' 11"
Eastern Spruce	2 x 10	14'	11' 4"
Eastern Spruce	2 x 12	17'	13' 10"
Table refers to floor joists with a live load of 30 pounds per square foot			
Eastern Spruce	2 x 6	9' 6"	7' 9"
Eastern Spruce	2 x 8	12' 6"	10' 2"
Eastern Spruce	2 x 10	15' 11"	13'
Eastern Spruce	2 x 12	19' 4"	15' 10"
Table refers to floor joists with a live load of 20 pounds per square foot			
Eastern Spruce	2 x 6	10' 10"	8' 10"
Eastern Spruce	2 x 8	14' 3"	11' 8"
Eastern Spruce	2 x 10	18' 3"	14' 11"

TABLE 4.13 ■ Acceptable spans for Grade Number 2 construction joists

Wood Species	Joist Size	16 inches on center	24 inches on center
Table refers to floor joists with a live load of 40 pounds per square foot			
Eastern Hemlock	2 x 6	8' 7"	7' 6"
Eastern Hemlock	2 x 8	11' 4"	9' 11"
Eastern Hemlock	2 x 10	14' 6"	12' 8"
Eastern Hemlock	2 x 12	17' 7"	15' 4"
Table refers to floor joists with a live load of 30 pounds per square foot			
Eastern Hemlock	2 x 6	9' 6"	8' 3"
Eastern Hemlock	2 x 8	12' 6"	10' 11"
Eastern Hemlock	2 x 10	15' 11"	13' 11"
Eastern Hemlock	2 x 12	19' 4"	16' 11"
Table refers to floor joists with a live load of 20 pounds per square foot			
Eastern Hemlock	2 x 6	10' 10"	9' 6"
Eastern Hemlock	2 x 8	14' 3"	12' 6"
Eastern Hemlock	2 x 10	18' 3"	15' 11"

TABLE 4.14 ■ Acceptable spans for Grade Number 2 construction joists

Wood Species	Joist Size	16 inches on center	24 inches on center
Table refers to floor joists with a live load of 40 pounds per square foot			
Douglas Fir-South	2 x 6	9' 1"	7' 11"
Douglas Fir-South	2 x 8	12'	10' 6"
Douglas Fir-South	2 x 10	15' 3"	13' 4"
Douglas Fir-South	2 x 12	18' 7"	16' 3"
Table refers to floor joists with a live load of 30 pounds per square foot			
Douglas Fir-South	2 x 6	10'	8' 9"
Douglas Fir-South	2 x 8	13' 2"	11' 6"
Douglas Fir-South	2 x 10	16' 10"	14' 8"
Douglas Fir-South	2 x 12	20' 6"	17' 11"
Table refers to floor joists with a live load of 20 pounds per square foot			
Douglas Fir-South	2 x 6	11' 5"	10'
Douglas Fir-South	2 x 8	15' 1"	13' 2"
Douglas Fir-South	2 x 10	19' 3"	16' 10"

TABLE 4.15 ■ Acceptable spans for Grade Number 2 construction joists

Wood Species	Joist Size	16 inches on center	24 inches on center
Table refers to floor joists with a live load of 40 pounds per square foot			
Douglas Fir-Larch	2 x 6	9' 11"	8' 6"
Douglas Fir-Larch	2 x 8	13' 1"	11' 3"
Douglas Fir-Larch	2 x 10	16' 9"	14' 4"
Douglas Fir-Larch	2 x 12	20' 4"	17' 5"
Table refers to floor joists with a live load of 30 pounds per square foot			
Douglas Fir-Larch	2 x 6	10' 9'	7"
Douglas Fir-Larch	2 x 8	14' 5"	12' 7"
Douglas Fir-Larch	2 x 10	18' 5"	16' 1"
Douglas Fir-Larch	2 x 12	22' 5"	19' 7"
Table refers to floor joists with a live load of 20 pounds per square foot			
Douglas Fir-Larch	2 x 6	12' 6"	10' 11"
Douglas Fir-Larch	2 x 8	16' 6"	14' 5"
Douglas Fir-Larch	2 x 10	21' 1"	18' 5"

TABLE 4.16 ■ **Acceptable spans for Grade Number 2 construction joists**

Wood Species	Joist Size	16 inches on center	24 inches on center
Table refers to floor joists with a live load of 40 pounds per square foot			
Balsam Fir	2 x 6	9' 1"	7' 9"
Balsam Fir	2 x 8	12'	10' 2"
Balsam Fir	2 x 10	15' 3"	13'
Balsam Fir	2 x 12	18' 7"	15' 10"
Table refers to floor joists with a live load of 30 pounds per square foot			
Balsam Fir	2 x 6	10'	8 '6"
Balsam Fir	2 x 8	13' 2"	11' 3"
Balsam Fir	2 x 10	16' 10"	14' 4"
Balsam Fir	2 x 12	20' 6"	17' 5"
Table refers to floor joists with a live load of 20 pounds per square foot			
Balsam Fir	2 x 6	11' 5"	10'
Balsam Fir	2 x 8	15' 1"	13' 2"
Balsam Fir	2 x 10	19' 3"	16' 10"

TABLE 4.17 ■ **Acceptable spans for Grade Number 2 construction joists**

Wood Species	Joist Size	16 inches on center	24 inches on center
Table refers to floor joists with a live load of 40 pounds per square foot			
California Redwood	2 x 6	8' 4"	7' 3"
California Redwood	2 x 8	11'	9' 7"
California Redwood	2 x 10	14'	12' 3"
California Redwood	2 x 12	17'	14' 11"
Table refers to floor joists with a live load of 30 pounds per square foot			
California Redwood	2 x 6	9' 10	8' 7"
California Redwood	2 x 8	13'	11' 4"
California Redwood	2 x 10	16' 7	14' 6"
California Redwood	2 x 12	20' 2	17' 8"
Table refers to floor joists with a live load of 20 pounds per square foot			
California Redwood	2 x 6	11' 4"	9' 10"
California Redwood	2 x 8	14' 10"	13'
California Redwood	2 x 10	19'	16' 7"

TABLE 4.18 ■ Acceptable spans for Grade Number 2 construction joists

Wood Species	Joist Size	16 inches on center	24 inches on center
Table refers to floor joists with a live load of 40 pounds per square foot			
Western Cedars	2 x 6	8' 4"	7' 3"
Western Cedars	2 x 8	11'	9' 7"
Western Cedars	2 x 10	14'	12' 3"
Western Cedars	2 x 12	17'	14' 11"
Table refers to floor joists with a live load of 30 pounds per square foot			
Western Cedars	2 x 6	9' 2"	8'
Western Cedars	2 x 8	12' 1"	10' 7"
Western Cedars	2 x 10	15' 5"	13' 6"
Western Cedars	2 x 12	18' 9"	16' 5"
Table refers to floor joists with a live load of 20 pounds per square foot			
Western Cedars	2 x 6	10' 6"	9' 2"
Western Cedars	2 x 8	13' 10"	12' 1"
Western Cedars	2 x 10	17' 8"	15' 5"

TABLE 4.19 ■ Acceptable spans for Grade Number 2 construction joists

Wood Species	Joist Size	16 inches on center	24 inches on center
Table refers to floor joists with a live load of 40 pounds per square foot			
Southern Pine	2 x 6	9' 9"	8' 4"
Southern Pine	2 x 8	12' 10"	11'
Southern Pine	2 x 10	16' 5"	14'
Southern Pine	2 x 12	19' 11"	17'
Table refers to floor joists with a live load of 30 pounds per square foot			
Southern Pine	2 x 6	10' 9"	9' 4"
Southern Pine	2 x 8	14' 2"	12' 4"
Southern Pine	2 x 10	18'	15' 9"
Southern Pine	2 x 12	21' 11"	19' 2"
Table refers to floor joists with a live load of 20 pounds per square foot			
Southern Pine	2 x 6	12' 3"	10' 9"
Southern Pine	2 x 8	16' 2"	14' 2"
Southern Pine	2 x 10	20' 8"	18'

TABLE 4.20 ▪ Acceptable spans for Grade Number 2 construction joists

Wood Species	Joist Size	16 inches on center	24 inches on center
Table refers to floor joists with a live load of 40 pounds per square foot			
Mountain Hemlock	2 x 6	8' 7"	7' 6"
Mountain Hemlock	2 x 8	11' 4"	9' 11"
Mountain Hemlock	2 x 10	14' 6"	12' 8"
Mountain Hemlock	2 x 12	17' 7"	15' 4"
Table refers to floor joists with a live load of 30 pounds per square foot			
Mountain Hemlock	2 x 6	9' 6"	8' 3"
Mountain Hemlock	2 x 8	12' 6"	10' 11"
Mountain Hemlock	2 x 10	15' 11"	13' 11"
Mountain Hemlock	2 x 12	19' 4"	16' 11"
Table refers to floor joists with a live load of 20 pounds per square foot			
Mountain Hemlock	2 x 6	10' 10"	9' 6"
Mountain Hemlock	2 x 8	14' 3"	12' 6"
Mountain Hemlock	2 x 10	18' 3"	15' 11"

TABLE 4.21 ▪ Acceptable spans for Grade Number 2 construction joists

Wood Species	Joist Size	16 inches on center	24 inches on center
Table refers to floor joists with a live load of 40 pounds per square foot			
Northern Pine	2 x 6	9' 1"	7' 3"
Northern Pine	2 x 8	12' 4"	9' 7"
Northern Pine	2 x 10	15' 3"	12' 3"
Northern Pine	2 x 12	18' 7"	14' 11"
Table refers to floor joists with a live load of 30 pounds per square foot			
Northern Pine	2 x 6	10'	8' 3"
Northern Pine	2 x 8	13' 2"	10' 11"
Northern Pine	2 x 10	16' 10"	13' 11"
Northern Pine	2 x 12	20' 6"	16' 11"
Table refers to floor joists with a live load of 20 pounds per square foot			
Northern Pine	2 x 6	11' 5"	9' 6"
Northern Pine	2 x 8	15' 1"	12' 6"
Northern Pine	2 x 10	19' 3"	15' 11"

TABLE 4.22 ■ Acceptable spans for Grade Number 2 construction joists made

Wood Species	Joist Size	16 inches on center	24 inches on center
Table refers to floor joists with a live load of 40 pounds per square foot			
Eastern White Pine	2 x 6	8' 4"	6' 9"
Eastern White Pine	2 x 8	11'	8' 11"
Eastern White Pine	2 x 10	14'	11' 4"
Eastern White Pine	2 x 12	17'	13' 10"
Table refers to floor joists with a live load of 30 pounds per square foot			
Eastern White Pine	2 x 6	9' 6"	7' 9"
Eastern White Pine	2 x 8	12' 6"	10' 2"
Eastern White Pine	2 x 10	15' 11"	13'
Eastern White Pine	2 x 12	19' 4"	15' 10"
Table refers to floor joists with a live load of 20 pounds per square foot			
Eastern White Pine	2 x 6	10' 10"	8' 10"
Eastern White Pine	2 x 8	14' 3"	11' 8"
Eastern White Pine	2 x 10	18' 3"	14' 11"

DON'T SKIMP

Don't skimp when sizing joists. Some contractors stretch the limits to save a few dollars. This is not smart business. No one wants to go back into a finished job and replace broken or sagging joists. If there is any doubt in your mind about a marginal situation, either shorten the joist span or enlarge the joist size.

chapter 5

FRAMING WALLS AND FLOORS

The framing of walls and floors is a big part of any major construction job. Experienced framers can usually do this work without the job appearing to be difficult. However, those who are not experienced in framing find the task daunting. Whether you are framing a small storage shed, a house, a barn, or a commercial building, the principles are essentially the same.

There was a time when metal studs were not usually seen in residential framing. Wood is still, by far, the leading choice of framing materials for homes, but metal studs are cutting into the framing phase of construction for residential purposes. Regional and personal preference has a lot to do with the choice of materials. Money also plays a part in the selection of building materials.

Chapter 4 outlines the selection and use of sills and joists. This is a part of framing, but here we will be discussing much more of the framing process. Our main focus in this chapter is floor structures, subflooring, and wall framing.

JOISTS AND BEAMS

We won't spend a lot of time on joists and beams in this chapter, since we talked about them in the last chapter. However, since they are a part of framing, we will touch on them here.

▶ *trade* **tip**

You have probably seen markings on plywood countless times. The odds are high that you have used C-D plywood. But, do you know what the letters mean? The "C" identifies the grade of the face veneer, and the "D" identifies the grade of the back veneer.

► *trade* **tip**

When building decks, use galvanized nails to reduce the risk of nails rusting and staining the decking material.

Decks

We have not talked about decks, so the following tables will help you choose joists for deck construction.

Decks and their floor structures are frequently constructed with materials that have been treated to resist rotting for many, many years. In the past, this material contained harmful chemicals that are now being phased out. Modern treated wood is said to be safer than what most carpenters used when they started in the trade.

Basic Floor Structure Connections

Basic floor structure connections are fairly simple. Whether you prefer to support joists using ledger strips or metal joist hangers, the important key is to make sure that the joists are supported. There are carpenters who feel that nailing joists into a girder is all that is needed. I've seen this type of construction countless times, and I can't recall a situation where the

TABLE 5.1 ▪ Common beam spans for decks

Beam size	Maximum distance between support post
4-x-6	6 feet
4-x-8	8 feet
4-x-10	10 feet
4-x-12	12 feet

TABLE 5.2 ▪ Common joist spans for decks where joists are Installed 16 inches on center

Beam size	Maximum span
2-x-6	8 feet
2-x-8	10 feet
2-x-10	13 feet

TABLE 5.3 ▪ Common joist spans for decks where joists are Installed 24 inches on center

Beam size	Maximum span
2-x-6	6 feet
2-x-8	8 feet
2-x-10	10 feet

TABLE 5.4 ▪ Common joist spans for decks where joists are Installed inches on center

Beam size	Maximum span
2-x-6	5 feet
2-x-8	7 feet
2-x-10	8 feet

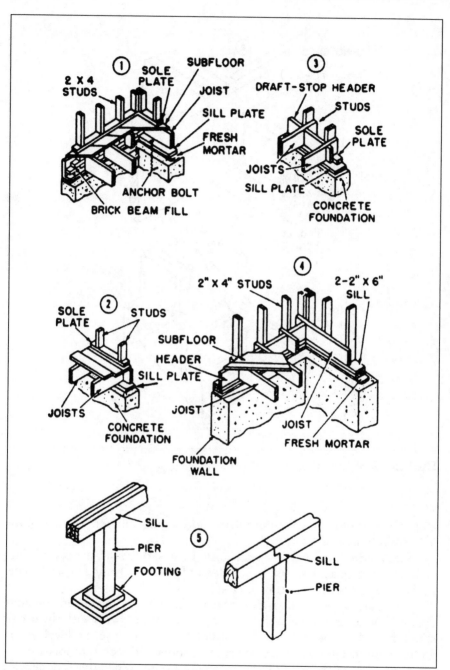

FIGURE 5.1 ■ Types of sills

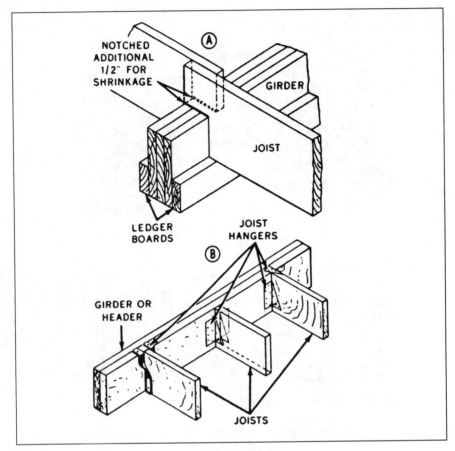

FIGURE 5.2 ■ Joist to girder attachment

connections failed. But, having a little more support under the ends of the joist is far better.

The methods used to connect joists vary with circumstances. Ceiling joists attached to a girder will look different from joists resting on a sill plate. In the case of floor joists sitting on a sill, the sill is the support.

There are times, such as adding an addition to a home, when you must deal with varying floor levels. One way of handling this is with the use of stairs. But when the difference is minimal, you can raise the level of the addition by installing a ledger stringer to support the ends of joists.

Many carpenters build their own beams on site. When this is done with wood, the joints of wood members should be staggered. This practice maintains greater strength for the beam.

Joists sometimes require reinforcement to support unusually heavy loads. This may be the case under a bearing wall. This process can involve doubling

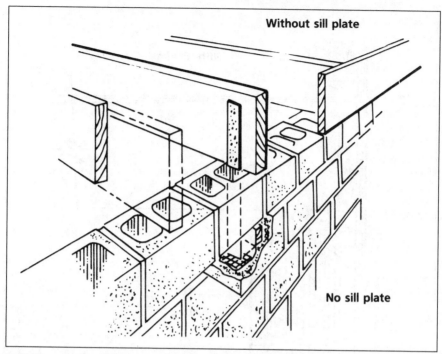

Without sill plate

No sill plate

FIGURE 5.3 ▪ **Floor joist detail without a sill plate**
(Drawing courtesy of USDA Forest Service)

joists and attaching them to each other, or placing them close to each other and installing blocking between the two joists.

If you have to build a joist system over a concrete floor, such as in a basement or garage, you can use a sleeper system. Screeds can be placed flat on the concrete if space is not needed below the floor. Or, you can stand joists up, as in normal construction, to create working space for mechanical systems.

SUBFLOORING

Once a flooring system is constructed, you are ready for subflooring. This is the rough flooring over which finish flooring will be installed at a later date. Plywood has long been a favorite for subflooring, but there are many other options, some of which include:

- CDX plywood
- Tongue-and-groove plywood
- Oriented-strand board (OSB) board
- Specialized, weather-resistant panels
- Wood planks

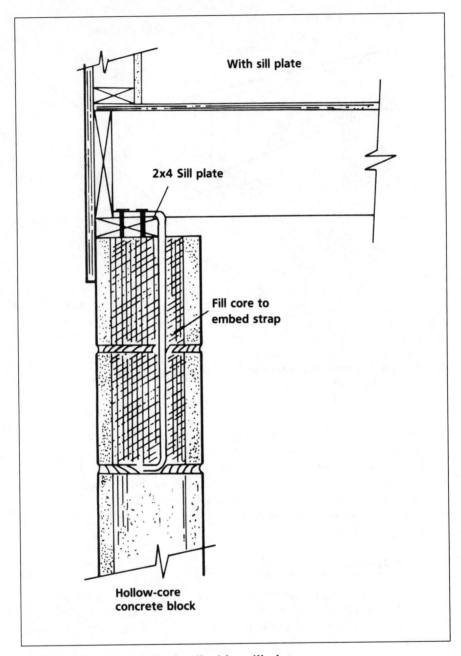

FIGURE 5.4 ▪ Floor joist detail with a sill plate
(Drawing courtesy of USDA Forest Service)

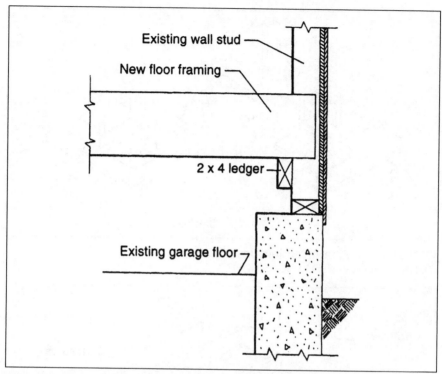

FIGURE 5.5 ■ Framing to bring a floor in an addition, existing porch, or garage up to the same floor level as a home

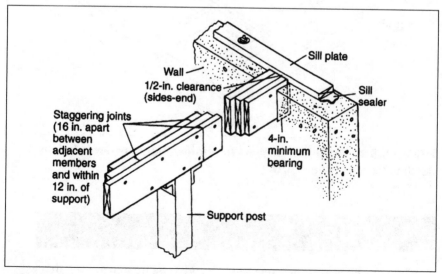

FIGURE 5.6 ■ Beam construction

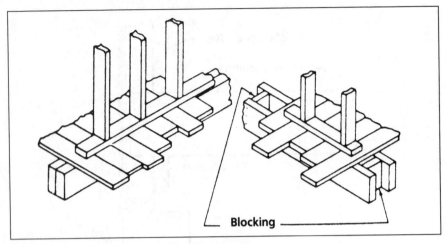

FIGURE 5.7 ■ Reinforced joist installation
(Drawing courtesy of USDA Forest Service)

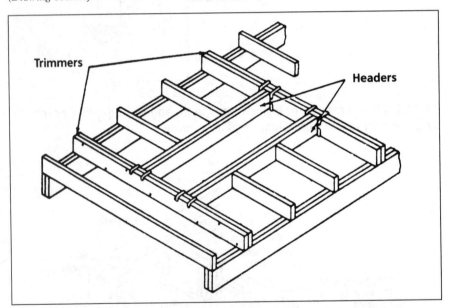

FIGURE 5.8 ■ The use of headers to reinforce joists where a hole is needed, such as for a stairway
(Drawing courtesy of USDA Forest Service)

► *trade* **tip**

When installing joists or screeds on concrete, use treated lumber that will not be damaged by the moisture transferred from the concrete.

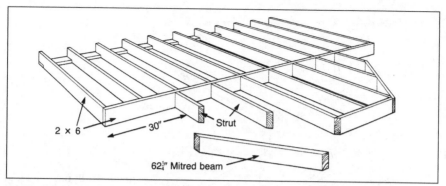

FIGURE 5.9 ■ **Sleeper system used to build over existing concrete**
(Drawing courtesy of Georgia-Pacific Corp.)

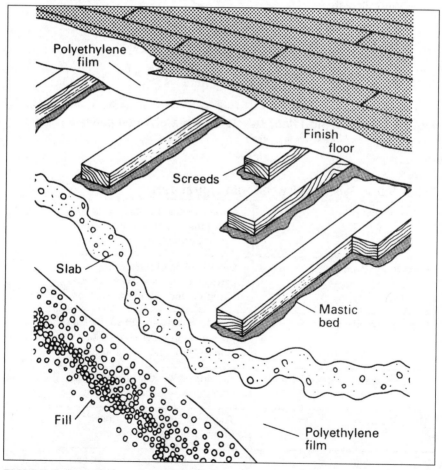

FIGURE 5.10 ■ **Using screeds to build a flooring system over existing concrete**
(Drawing courtesy of U.S. Dept. of Agriculture)

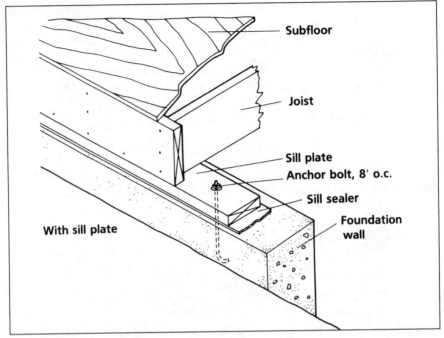

FIGURE 5.11 ■ Typical floor structure with a plywood panel
(Chart courtesy of USDA Forest Service)

TABLE 5.5 ■ Types of plywood with typical uses

Softwood veneer	Cross laminated plies or veneers—Sheathing, general construction and industrial use, etc.
Hardwood veneer	Cross laminated plies with hardwood face and back veneer—Furniture and cabinet work, etc.
Lumbercore plywood	Two face veneers and 2 crossband plies with an inner core of lumber strips—Desk and table tops, etc.
Medium-density overlay (MDO)	Exterior plywood with resin and fiber veneer—Signs, soffits, etc.
High-density overlay (HDO)	Tougher than MDO—Concrete forms, workbench tops, etc.
Plywood siding	T-111 and other textures used as one step sheathing and siding where codes allow.

☑ *fast***facts**

Most codes require two layers of subflooring if the material used is some type of panel that does not use tongue-and-groove (T&G) construction. By using a thicker, T&G material, such as ¾-inch T&G plywood, you can eliminate one layer of subflooring. This can be a real timesaver.

TABLE 5.6 ■ Reconstituted wood panels with typical uses

Particleboard	Wood particles and resin. a. Industrial grade—Cabinets and counter tops under plastic laminates. b. Underlayment—Installed over subfloor under tile or carpet.
Wafer board	Wood wafers and resin. Inexpensive sheathing, craft projects, etc.
Oriented Strand Board (OSB)	Thin wood strands oriented at right angles with phenolic resin. Same uses as above.
Hardboard	Wood fiber mat compressed into stiff, hard sheets. a. Service—Light weight—Cabinet backs, etc. b. Standard—Stronger with better finish quality. c. Tempered—Stiffer and harder for exterior use.
Fiberboard (Grayboard)	Molded wood fibers—Underlayment, sound deadening panels.
Composite plywood	A core of particleboard with face and back veneers glued directly to it.

FASTENERS

Fasteners, whether nails or screws, are needed to frame a structure. Adhesives are also used in many cases, especially when installing subflooring. With today's wide selection of nails, screws, pneumatic nails, staples, and brads, there is no shortage of options. Match your fasteners to your task and always follow manufacturer recommendations.

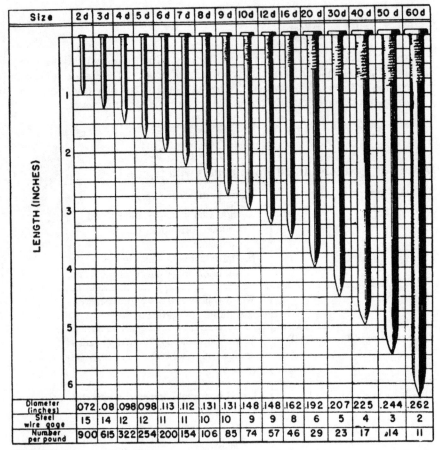

Size	2 d	3 d	4 d	5 d	6 d	7 d	8 d	9 d	10d	12d	16 d	20 d	30d	40 d	50 d	60 d
Diameter (inches)	.072	.08	.098	.098	.113	.112	.131	.131	.148	.148	.162	.192	.207	225	.244	.262
Steel wire gage	15	14	12	12	11	11	10	10	9	9	8	6	5	4	3	2
Number per pound	900	615	322	254	200	154	106	85	74	57	46	29	23	17	14	11

FIGURE 5.12 ■ Nail sizes
(Chart courtesy of USDA Forest Service)

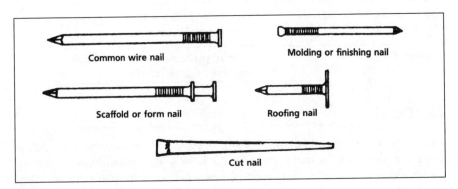

FIGURE 5.13 ■ Types of nails
(Drawing courtesy of USDA Forest Service)

TABLE 5.7 ■ **Recommended uses of nails**

Size	Lgth (in.)	Diam (in.)	Remarks	Where used
2d	1	.072	Small head	Finish work, shop work.
2d	1	.072	Large flathead	Small timber, wood shingles, lathes.
3d	1¼	.08	Small head	Finish work, shop work.
3d	1¼	.08	Large flathead	Small timber, wood shingles, lathes.
4d	1½	.098	Small head	Finish work, shop work.
4d	1½	.098	Large flathead	Small timber, lathes, shop work.
5d	1¾	.098	Small head	Finish work, shop work.
5d	1¾	.098	Large flathead	Small timber, lathes, shop work.
6d	2	.113	Small head	Finish work, casing, stops, etc., shop work.
6d	2	.113	Large flathead	Small timber, siding, sheathing, etc., shop work.
7d	2¼	.113	Small head	Casing, base, ceiling, stops, etc.
7d	2¼	.113	Large flathead	Sheathing, siding, subflooring, light framing.
8d	2½	.131	Small head	Casing, base, ceiling, wainscot, etc., shop work.
8d	2½	.131	Large flathead	Sheathing, siding, subflooring, light framing, shop work.
8d	1¼	.131	Extra-large flathead	Roll roofing, composition shingles.
9d	2¾	.131	Small head	Casing, base, ceiling, etc.
9d	2¾	.131	Large flathead	Sheathing, siding, subflooring, framing, shop work.
10d	3	.148	Small head	Casing, base, ceiling, etc., shop work.
10d	3	.148	Large flathead	Sheathing, siding, subflooring, framing, shop work.
12d	3¼	.148	Large flathead	Sheathing, subflooring, framing.
16d	3½	.162	Large flathead	Framing, bridges, etc.
20d	4	.192	Large flathead	Framing, bridges, etc.
30d	4½	.207	Large flathead	Heavy framing, bridges, etc.
40d	5	.225	Large flathead	Heavy framing, bridges, etc.
50d	5½	.244	Large flathead	Extra-heavy framing, bridges, etc.
60d	6	.262	Large flathead	Extra-heavy framing, bridges, etc.

[1] This chart applies to wire nails, although it may be used to determine the length of cut nails.
(Chart courtesy of USDA Forest Service)

TABLE 5.8 ■ Nail sizes and the approximate number of nails per pound

Penny size "d"	Length	Approximate number per pound, Common	Approximate number per pound, Box	Approximate number per pound, Finish
2	1"	875	1000	1300
3	1¼"	575	650	850
4	1½"	315	450	600
5	1¾"	265	400	500
6	2"	190	225	300
7	2¼"	160		
8	2½"	105	140	200
9	2¾"	90		
10	3"	70	90	120
12	3¼"	60	85	110
16	3½"	50	70	90
20	4"	30	50	60
30	4½"	25		
40	5"	20		
50	5½"	15		
60	6"	10		

NOTE: Aluminum and c.c. nails are slightly smaller than other nails of the same penny size.

Pneumatic nail guns can reduce labor time tremendously. Most production carpenters today do rely on pneumatics. The days of swinging a framing hammer are few and far between for carpenters who are working on fixed-price jobs.

FAST MATH

Who doesn't like fast math when it comes to doing take-offs? How would you estimate the number of joists needed for a house that is 40 feet in length?

TABLE 5.9 ■ Screw lengths and available gauge numbers

Length	Gauge numbers	Length	Gauge numbers
¼"	0 to 3	1¾"	8 to 20
⅜"	2 to 7	2"	8 to 20
½"	2 to 8	2¼"	9 to 20
⅝"	3 to 10	2½"	12 to 20
¾"	4 to 11	2¾"	14 to 20
⅞"	6 to 12	3"	16 to 20
1"	6 to 14	3½"	18 to 20
1¼"	7 to 16	4"	18 to 20
1½"	6 to 18		

▶ *trade* **tip**

When nailing subflooring panels, edges should have nails spaced about 6 inches apart. Interior nailing can be spaced up to 10 inches apart.

Want to do it the easy way? If so, multiply the length of the structure, in this case 40 feet, by ¾. This is based on the assumption that the joists will be installed 16 inches on center. When you do the math, you come up with a need for 30 joists, if each joist spans the entire width of the house. Then add one more joist and you have a pretty good take-off. If you have a wider house and will need joists to run on each side of the girder to cover the span, double your finding for a total of 62 joists. This formula will also work for rafters and studs. You will, of course, have to add in more material for blocking, headers, and so forth. If you are calculating studs, you will have to factor in additional studs for jacks, cripples, and so forth. But it is fast math that works.

☑ *fast***facts**

Plates and shoes for walls can be estimated by taking the perimeter footage of a building and multiplying it by 3.

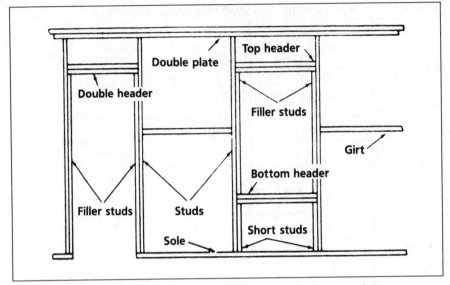

FIGURE 5.14 ■ **Typical door and window framing that requires additional material during the framing process**
(Drawing courtesy of USDA Forest Service)

TABLE 5.10 ■ Converting fractions to decimals

Fractions	Decimal equivalent	Fractions	Decimal equivalent
1/16	0.0625	9/16	0.5625
1/8	0.1250	5/8	0.6250
3/16	0.1875	11/16	0.6875
1/4	0.2500	3/4	0.7500
5/16	0.3125	13/16	0.8125
3/8	0.3750	7/8	0.8750
7/16	0.4375	15/16	0.9375
1/2	0.5000	1	1.000

If you are working with a building where structural members are placed on 24-inch centers, you can use a formula similar to the one used when building 16 inches on center. To do this, take the length of the building and multiply it by ½. Remember to add one additional member and to allow for the extra materials you will need for headers, jacks, blocking, and so forth. When figuring rafters with either formula, remember to double your answer to allow for rafters on both sides of the ridge pole.

Conversion tables are always helpful when looking for shortcuts to doing math. Whether you have to convert fractions to decimals or inches to millimeters, a conversion table makes the math much easier.

If you like these tables, you will want to check out the appendix at the end of this book. It is full of helpful tables, formulas, and data for all sorts of needs.

TABLE 5.11 ■ Converting inches to millimeters

Inches	Millimetres	Inches	Millimetres
1	25.4	11	279.4
2	50.8	12	304.8
3	76.2	13	330.2
4	101.6	14	355.6
5	127.0	15	381.0
6	152.4	16	406.4
7	177.8	17	431.8
8	203.2	18	457.2
9	228.6	19	482.6
10	254.0	20	508.0

TABLE 5.12 ■ Converting square inches to square centimeters

Square inches	Square centimeters	Square inches	Square centimeters
1	6.5	8	52.0
2	13.0	9	58.5
3	19.5	10	65.0
4	26.0	25	162.5
5	32.5	50	325.0
6	39.0	100	650.0
7	45.5		

TABLE 5.13 ■ Converting square feet to square meters

Square feet	Square meters	Square feet	Square meters
1	0.925	8	0.7400
2	0.1850	9	0.8325
3	0.2775	10	0.9250
4	0.3700	25	2.315
5	0.4650	50	4.65
6	0.5550	100	9.25
7	0.6475		

TABLE 5.14 ▪ Formula functions

Circle

Circumference = diameter × 3.1416

Circumference = radius × 6.2832

Diameter = radius × 2

Diameter = square root of; (area ÷ 0.7854)

Diameter = square root of area × 1.1283

Diameter = circumference × 0.31831

Radius = diameter ÷ 2

Radius = circumference × 0.15915

Radius = square root of area × 0.56419

Area = diameter × diameter × 0.7854

Area = half of the circumference × half of the diameter

Area = square of the circumference × 0.0796

Arc length = degrees × radius × 0.01745

Degrees of arc = length ÷ (radius × 0.01745)

Radius of arc = length ÷ (degrees × 0.01745)

Side of equal square = diameter × 0.8862

Side of inscribed square = diameter × 0.7071

Area of sector = area of circle × degrees of arc ÷ 360

Cone

Area of surface = one half of circumference of base × slant height + area of base.

Volume = diameter × diameter × 0.7854 × one-third of the altitude.

(continued)

TABLE 5.14 ■ Formula functions *(continued)*

Cube

Volume = width × height × length

Cylinder

Area of surface = diameter × 3.1416 × length + area of the two bases

Area of base = diameter × diameter × 0.7854

Area of base = volume ÷ length

Length = volume ÷ area of base

Volume = length × area of base

Capacity in gallons = volume in inches ÷ 231

Capacity of gallons = diameter × diameter × length × 0.0034

Capacity in gallons = volume in feet × 7.48

Ellipse

Area = short diameter × long diameter × 0.7854

Hexagon

Area = width of side × 2.598 × width of side

Parallelogram

Area = base × distance between the two parallel sides

Pyramid

Area = ½ perimeter of base × slant height + area of base

Volume = area of base × ⅓ of the altitude

Rectangle

Area = length × width

Rectangular prism

Volume = width × height × length

TABLE 5.14 ■ Formula functions *(continued)*

Sphere
Area of surface = diameter × diameter × 3.1416
Side of inscribed cube = radius × 1.547
Volume = diameter × diameter × diameter × 0.5236
Square
Area = length × width
Triangle
Area = one-half of height times base
Trapezoid
Area = one-half of the sum of the parallel sides × the height

TABLE 5.15 ■ Board lumber conversions

Nominal size	Actual size	Board feet per linear foot	Linear feet per 1000 board feet
1 × 2	¾ × 1½	⅙ (0.167)	6000
1 × 3	¾ × 2½	¼ (0.250)	4000
1 × 4	¾ × 3½	⅓ (0.333)	3000
1 × 6	¾ × 5½	½ (0.500)	2000
1 × 8	¾ × 7¼	⅔ (0.666)	1500
1 × 10	¾ × 9¼	⅚ (0.833)	1200
1 × 12	¾ × 11¼	1 (1.0)	1000

TABLE 5.16 ■ Board measures

Nominal size	Actual size	Board feet per linear foot	Linear feet per 1000 board feet
2 × 2	1½ × 1½	⅓ (0.333)	3000
2 × 3	1½ × 2½	½ (0.500)	2000
2 × 4	1½ × 3½	⅔ (0.666)	1500
2 × 6	1½ × 5½	1 (1.0)	1000
2 × 8	1½ × 7¼	1⅓ (1.333)	750
2 × 10	1½ × 9¼	1⅔ (1.666)	600
2 × 12	1½ × 11¼	2 (2.0)	500

ROUGH OPENINGS

When you are framing walls, you have to create rough openings for windows and doors. It is best to have actual specifications for the windows and doors that will be used. These specs will give you dimensions for rough openings. Unfortunately, your customer may not know exactly what is going to be installed later. This forces you to do your own math for the rough openings.

As a rule of thumb, you should frame rough openings $2\frac{1}{8}$ inches wider than the finished product will be. The height should be 2 5/8 inches taller than the expected finish product. Remember, these are generic numbers. Get specific rough-in figures whenever possible. And, when in doubt, frame a

TABLE 5.17 ■ Header spans for headers with two wood members that have ½-inch plywood between the two header members

Lumber Size	Maximum Span
2 x 4	3 Feet
2 x 6	5 Feet
2 x 8	7 Feet
2 x 10	8 Feet
2 x 12	9 Feet

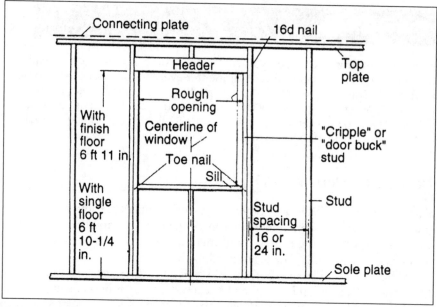

FIGURE 5.15 ■ Typical framing with a header over the rough opening for a window

larger hole. It is much easier to close in an opening than it is to expand it. I can give you some more general guidelines to work with.

- Double-hung window: rough opening width is the width of the glass plus 6 inches. The height for the rough opening is the height of the glass plus 6 inches.
- Casement window with double sash: rough opening for width is the width of the glass plus 1¼ inch. The opening for height is the height of the glass plus 6⅜ inches.
- Doors: rough opening for the width of the door should be the width of the door plus 2½ inches. The rough opening for height is the height of the door plus 3 inches.

☑ *fast***facts**

Top plates on walls serve many functions. They tie studs together, provide a nailing surface for wall coverings, support the lower ends of rafters, and establish a connection between walls and roofs.

▶ *trade* **tip**

If you are using 2 x 4 studs framed 16 inches on center for exterior walls, you will need 1.05 board feet of lumber for every square foot of area to be framed. This includes studs for corner bracing. You will also need 22 pounds of nails for every 1,000 board feet you install. If you are using 2 x 6 studs on 16-inch centers, you will need 1.51 board feet for every square foot framed and 15 pounds of nails for every 1,000 board feet installed.

FRAMING EXTERIOR WALLS

Framing exterior walls is not complicated work. There are some basics to follow and most of the work goes smoothly. A majority of carpenters build their walls on the subfloor and then stand them up. The walls have to be braced into place and the sole plate is attached to the floor system.

When stud walls are stood up and braced, they must be straightened. Once the walls are plumb, they can be secured into their permanent position. The braces will stay in place until enough of the framing is completed to keep the exterior walls from moving.

When framing walls, you have to frame openings for windows and doors. We have talked about establishing measurements for rough openings. As a reminder, try to obtain accurate rough-in dimensions for the actual products to be installed.

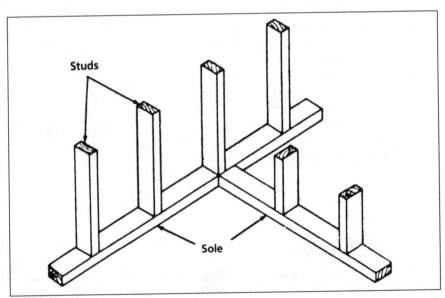

FIGURE 5.16 ■ The sole plate is attached to the floor system
(Drawing courtesy of USDA Forest Service)

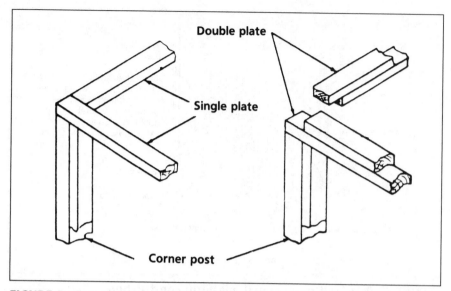

FIGURE 5.17 ■ Examples of a single top plate and a double top plate
(Drawing courtesy of USDA Forest Service)

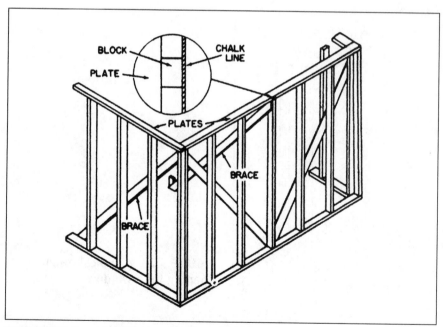

FIGURE 5.18 ■ Wall braces for exterior stud walls

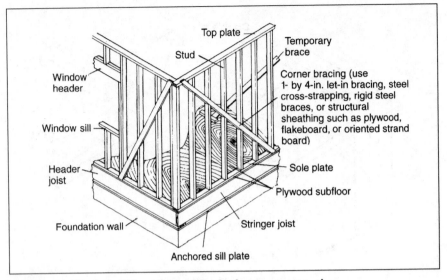

FIGURE 5.19 ■ Wall framing with platform construction

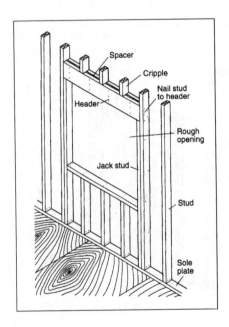

FIGURE 5.20 ■ Typical rough opening for a window

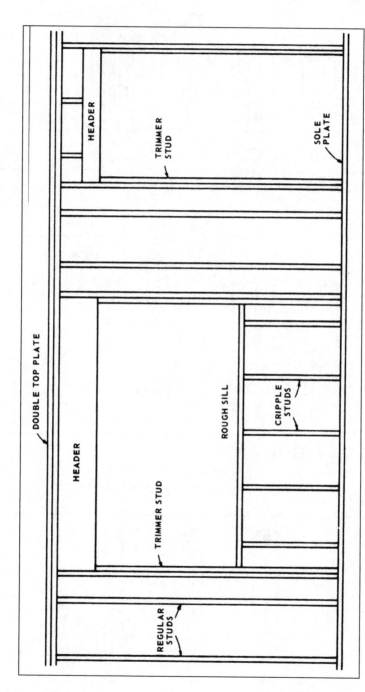

FIGURE 5.21 ■ Typical rough opening for a door and window

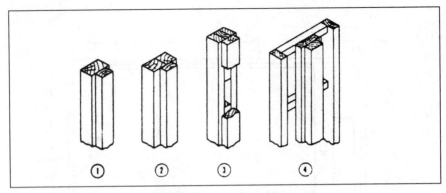

FIGURE 5.22 ■ T-post connectors

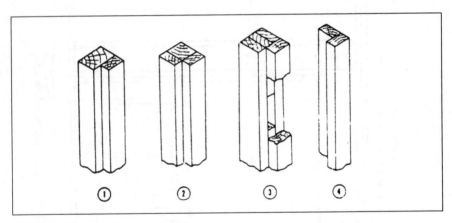

FIGURE 5.23 ■ Corner posts

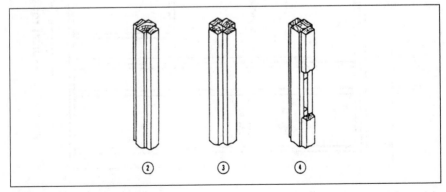

FIGURE 5.24 ■ Partition posts

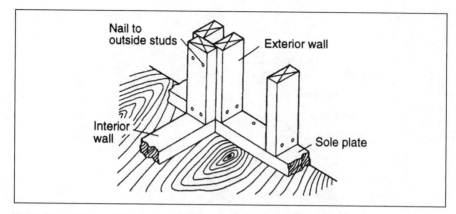

FIGURE 5.25 ▪ Tying a partition wall to an exterior wall

PARTITION WALLS

Partition walls have to be tied to other wall sections. There are many different ways to anchor partition walls to adjoining wall sections. You can use T-posts to tie walls together. Corner posts and partition posts are also used for this purpose. Remember to allow for these material needs when figuring a take-off for a job.

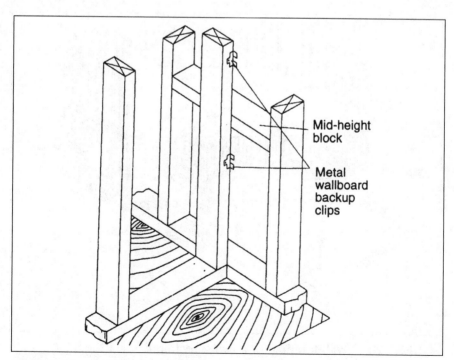

FIGURE 5.26 ▪ Using blocking as a means for attaching wall sections

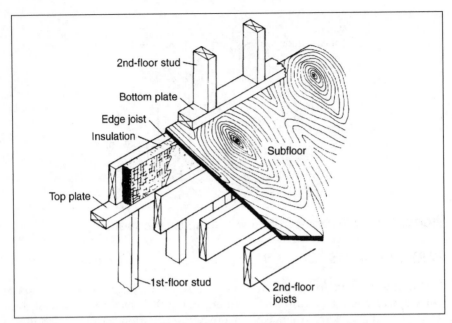

FIGURE 5.27 ■ Second-story framing for platform construction

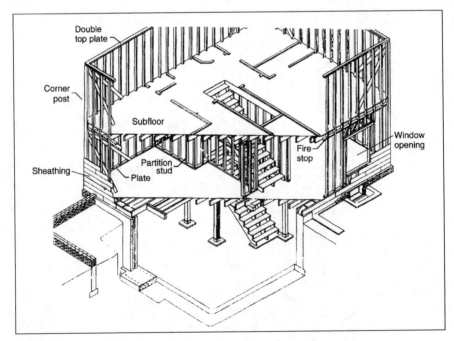

FIGURE 5.28 ■ Wall framing for a two-story home

MULTI-STORY WALLS

Multi-story walls are built in essentially the same fashion as single-story walls. The bottom plate of multi-story exterior walls are attached to edge joists and braced into place until enough framing is completed to hold the walls in their true position.

FULL FRAMING

Full framing is a major portion of a construction project. The framing has to be right for the rest of the job to work out. Walls have to be plumb. Backing for plumbing fixtures and similar items must be positioned properly to avoid problems when fixtures are set near the end of a job.

We will discuss other framing elements, such as stair construction and roof construction in upcoming chapters. The key is to get your measurements right and to maintain accuracy in your work. Framing is the skeleton of a building, and it must be done properly if the appearance of the work that follows is to show well.

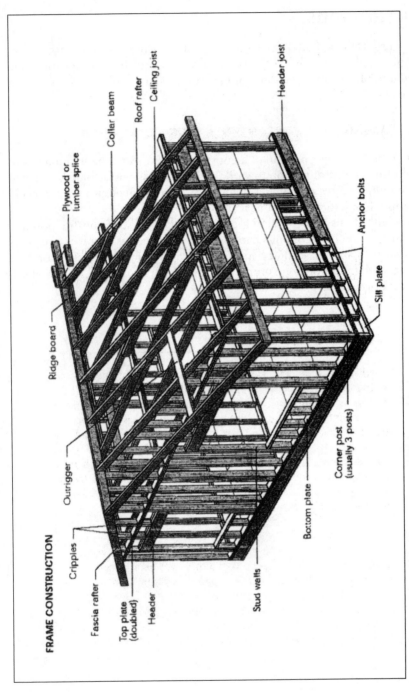

FIGURE 5.29 ■ Typical residential framing layout

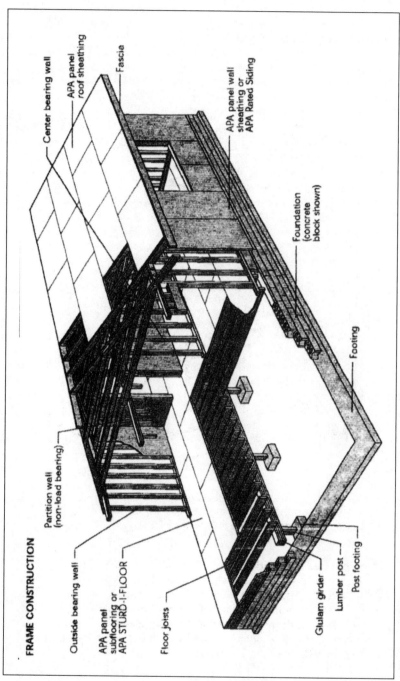

FRAME CONSTRUCTION

Partition wall (non-load bearing)

Outside bearing wall

APA panel subflooring or APA STURD-I-FLOOR

Floor joists

Glulam girder

Lumber post

Post footing

Center bearing wall

APA panel roof sheathing

Fascia

APA panel wall sheathing or APA Rated Siding

Foundation (concrete block shown)

Footing

FIGURE 5.30 ▪ Typical framing construction when using engineered wood products

chapter 6

STAIRS

S tairs are a standard part of many construction jobs. They are usually taken for granted, but they should not be. A set of stairs that does not have the proper rise or tread can present a serious danger to anyone using them. Some stairs are far too steep. Others have narrow treads that can make using them difficult, if not dangerous.

Experienced carpenters can take a framing square and cut risers and treads without much thought. But, many workers struggle with building stairs. Code requirements in local regions dictate the guidelines for proper stair construction. As always, consult your local building code to see what the minimum and maximum limits are for stair construction in your area.

FIGURING STAIR RISE

When figuring stair rise, you must consider the total vertical rise of the stair system. Most stairs are built once subflooring has been installed and prior to the installation of final floor coverings. Failing to take final flooring into account during the measuring phase can result in a stairway that is not built properly. And, you might be surprised at the number of carpenters who do forget to factor in the thickness for finished flooring.

▶ *trade* **tip**

When calculating the total vertical rise of stairs, you must factor in the thickness of any finished flooring to be installed. Failure to allow for the finished flooring will result in a stairway that does not have the proper spacing.

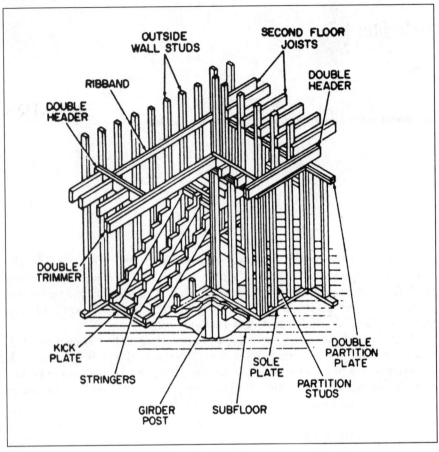

FIGURE 6.1 ■ Details of a stair system

The average unit rise for a set of stairs should be approximately 7 inches. If you total the unit rise and the unit run, the sum should be about 17.5 inches. It is a rule of thumb that the total vertical rise for a typical stairway will be about 8.5 feet. This would convert to 102 inches.

If you consider the figures given as averages, we can do the math to show you how to figure a set of stairs. You know that the vertical rise is 102 inches.

☑ *fast***facts**

To find the total run of a stairway, you must multiply the number of whole treads times the unit run.

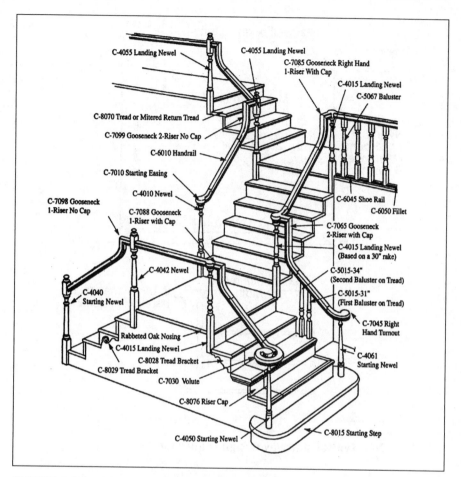

FIGURE 6.2 ■ Anatomy of a traditional staircase

How many risers will you need? The average unit rise is 7 inches, so you divide 102 inches by 7 inches and arrive an answer of 14.57 inches. When figuring stairs, eliminate the fraction of an answer. So, the number of risers needed is 14. Now you need to find the right number for the unit rise. To do this, divide 102 inches by 14. This will give you the answer of 7.29 inches, or about 7.25 inches. At this point, subtract the unit rise, which is 7.29 inches from the 17.5 inches that is assumed to be the sum of the unit rise and the unit run. Here is what the formula will look like: 17.5 − 7.29 = 10.21, or roughly 10.25 inches. The final result is a unit rise of about 7.25 inches, a unit run of about 10.25 inches, and a total of about 17.5 inches. This process is not difficult once you do it yourself on paper a few times.

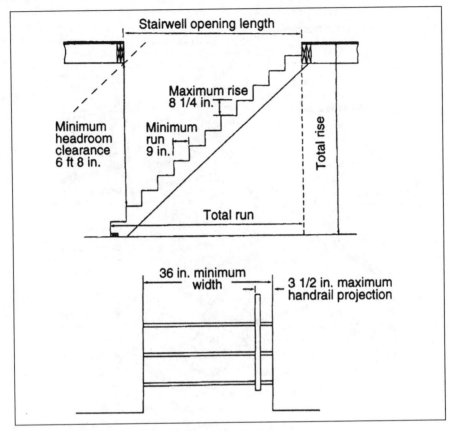

FIGURE 6.3 ■ Typical stairway design requirements

CALCULATING THE RUN

Calculating the run of a stairway requires multiplying the total number of whole treads times the unit run. It is common for the total number of whole treads to vary, depending upon how the upper end of the stairway is secured. Sometimes there will be a tread that is equal to the finished floor level. There are other times when the finished floor is treated as the last tread.

Using our previous example, assume that there are 14 whole treads in the staircase that we are working with. You know from the example that the

▶ *trade* **tip**

Keep a calculator and spare batteries for it in your truck. This simple tool will be very helpful when designing a stairway.

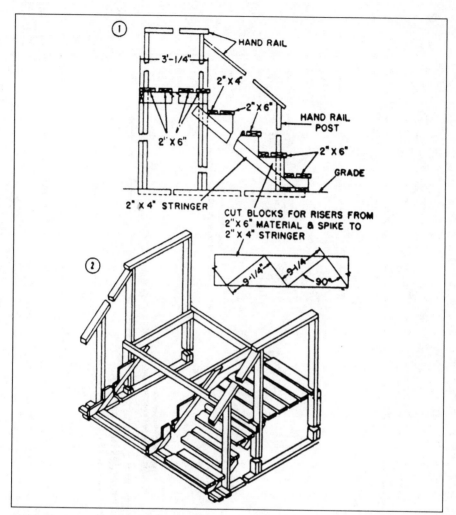

FIGURE 6.4 ■ Typical stair detail

unit run is 10.25. To calculate the total run of this stairway, multiply 14 by 10.25. You will arrive at a sum of 143.5 inches, or about 12 feet. Now, let's see how this would work if the finished floor at the top of the stairs was treated as a tread.

When the finished floor is used as the last tread at the top of a set of stairs, finding the total run of the stairway is done slightly differently. Instead of having 14 whole treads, you will have 13. Remember that the finished floor is taking the place of the last tread. This is why the number is reduced by one. Aside from this, the math is the same. Multiply 13 times 10.25 to arrive at an answer of 133.25 inches, or about 11 feet and 1 inch.

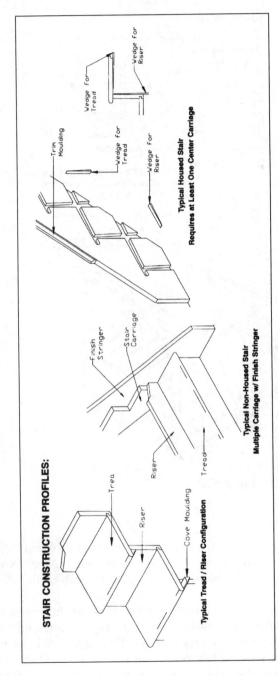

FIGURE 6.5 ■ Conventional stair profiles and parts
(By permission-Woodwork Institute, West Sacramento, Ca.)

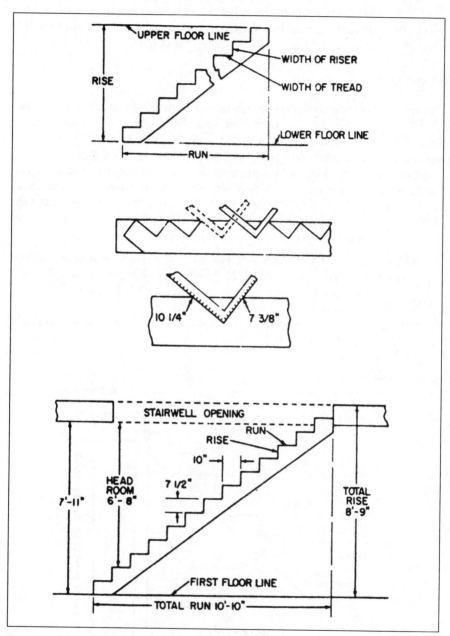

FIGURE 6.6 ■ Principal parts of stair construction

There may be times when there is a partial top tread. When this is the case, you add the actual run of the partial tread to your equation. Assume that you have a partial tread that has a rise of 5 inches. The formula will look like this: 13 x 10.25 = 133.25 + 5 = 11.52 inches, or about 11.5 feet.

STRINGERS

Figuring and cutting stringers is a substantial element of a successful staircase. The first step in the process is determining the length of the stock to work with. Figuring stringers can be a little tricky, but it is something that, once you do it a few times, you should not have a problem with the process.

You will be working with square roots when calculating stair stringers. Using our existing example, we know that the vertical rise is 8.5 feet and that the total run is 12 feet. To determine the length of a stringer, you have to take the vertical rise, square it, and multiply it by the total run, squared. Are you confused yet? Don't worry, it's not that hard. Here is what you do:

- To find the square of 8.5 feet, you can use a conversion table, such as one found in the appendix of this book. But to keep things simple, multiply 8.5 times 8.5 to arrive at an answer of 72.25. This is the square of 8.5. Not so hard, huh?
- Finding the square of the total run is done the same way. Multiply 12 times 12 to get 144.

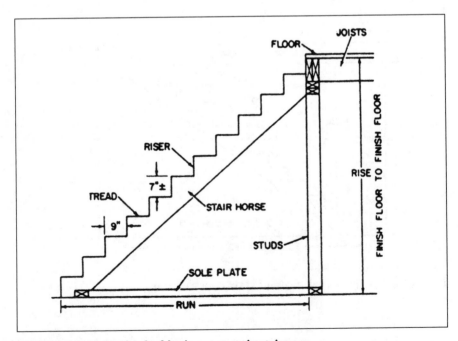

FIGURE 6.7 ■ Method of laying out stair stringers

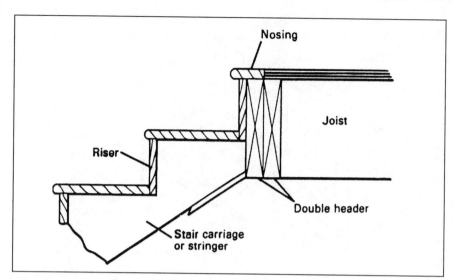

FIGURE 6.8 ■ Typical stair stringer

- Once you have the two numbers squared, you add them together. 72.25 + 144 equals 216.25. The square root of 216.25 is 14.71. This means that your stringer length will be about 14 feet, 5 inches long, or 173 inches.

TREAD THICKNESS

Tread thickness is a consideration when measuring the cuts on stringers. As you probably know, you locate the unit run on the tongue of a framing square and the unit rise on the body of the square. This allows you to mark the proper angles for cutting stringers. However, you must take into account the thickness of the treads to calculate all of your measurements accurately.

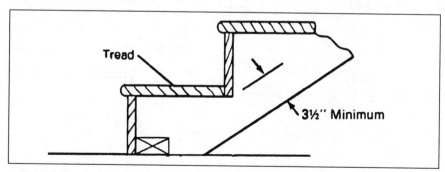

FIGURE 6.9 ■ Typical tread diagram

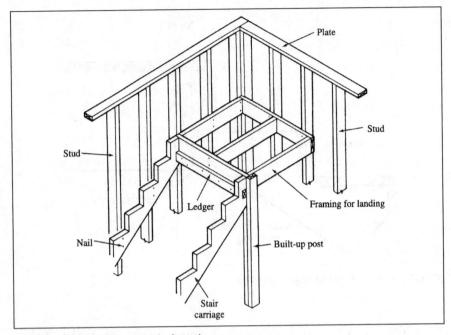

FIGURE 6.10 ■ Basic stair framing

Have you ever heard a term called "dropping the stringer"? This is the process of making an allowance for the thickness of treads. In general terms, if the stringer will sit on subflooring, you cut the stringer so that the first riser, and only the first riser, has a unit rise that is decreased by the thickness of one tread and the thickness of future finished flooring. When

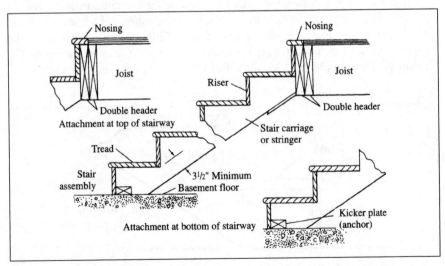

FIGURE 6.11 ■ Methods of attaching the base of stairs to a floor

☑ *fast*facts

Get to know your framing square well. If you are not well versed in the use of the square, invest some time in learning more about what you can do with it. While a square is a simple tool, it can enable you to accomplish many goals once you know the ins and outs of using the square.

the string is to sit on a finished floor, you reduce the first riser's unit rise by the thickness of one tread only. So, for example, if you have a unit rise of 7.25 inches and the stringer will sit on a finished floor and each tread has a thickness of one-eighth of an inch, you cut the unit rise at one-eighth of an inch less than 7.25 inches.

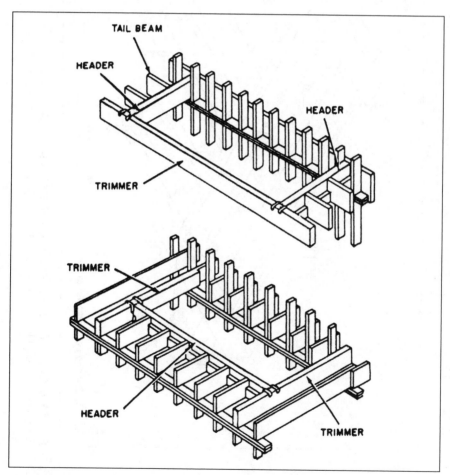

FIGURE 6.12 ■ Heading off a floor opening for a stairway

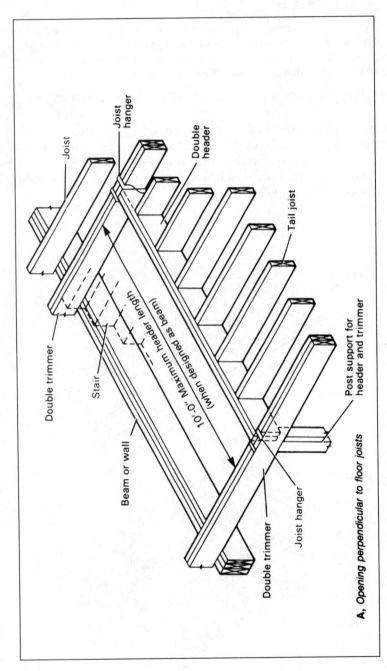

FIGURE 6.13A ■ **Framing details for stairway openings**
(Drawings courtesy of U.S. Dept. of Agriculture)

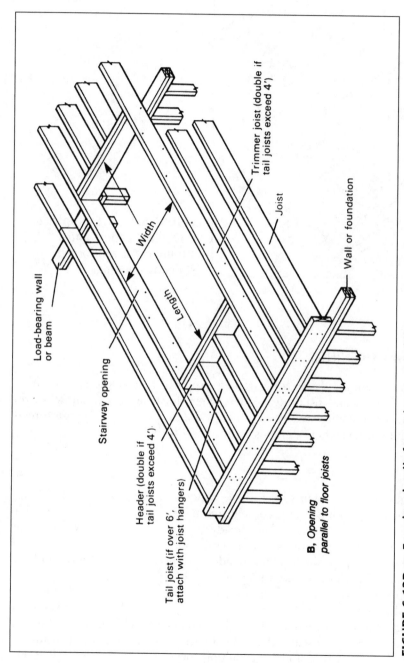

Load-bearing wall or beam

Stairway opening

Header (double if tail joists exceed 4')

Tail joist (if over 6', attach with joist hangers)

Width

Length

Trimmer joist (double if tail joists exceed 4')

Joist

Wall or foundation

B, Opening parallel to floor joists

FIGURE 6.13B ▪ **Framing details for stairway openings**
(Drawings courtesy of U.S. Dept. of Agriculture)

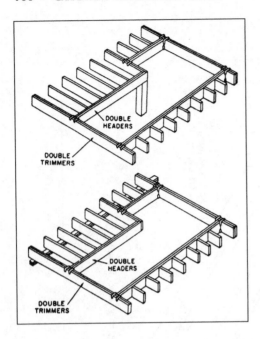

FIGURE 6.14 ■ Examples of doubling headers and trimmers for reinforced stairway openings.

STAIRWAY OPENINGS

Stairway openings have to be framed properly to provide structural integrity. This means installing headers where joists are cut to allow for the installation of a stairway. In some cases, the headers might be reinforced. Don't cut corners when it comes to heading off the openings.

WINDERS

It is not unusual for stairways to have winders in them. When winders are used, there are landings. This complicates matters a bit, but the basic math is done the same way as what has already been described. For example, you would base your calculations from the subflooring to a landing on the same criteria already described. When winders are used, there is more work involved, but the principles are the same.

Stairways are not as complicated as some carpenters feel they are. Certainly, some are more trying than others. Semi-spiral stairs can keep you thinking. Stairways with multiple levels can seem intimidating, but there is no reason for them to be scary. If you treat each set of stringers and treads the same, it is just a matter of making multiple measurements for multiple levels. All in all, stairs are not a big problem. But as a reminder, get to know your framing square inside out. It will serve you well.

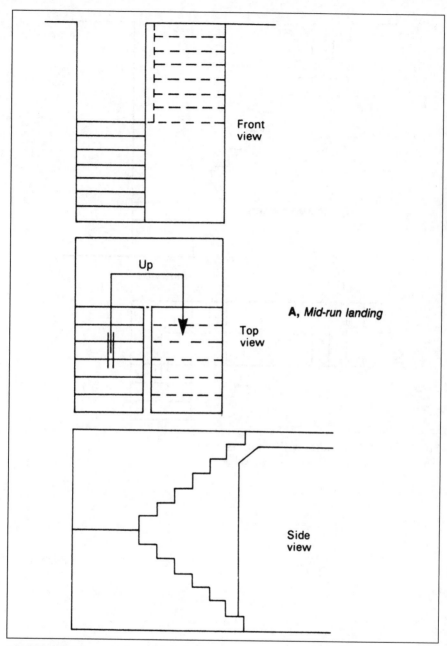

FIGURE 6.15A ■ Stairs with winders

(continued on next page)

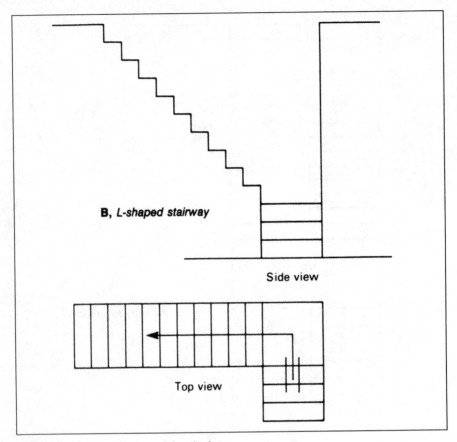

B, *L-shaped stairway*

Side view

Top view

FIGURE 6.15B ■ **Stairs with winders**
(Drawings courtesy of U.S. Dept. of Agriculture)

chapter 7

RAFTERS

C arpenters have been using rafters to build roof structures for years and years. While many builders today are using engineered trusses as a faster alternative to rafters, the use of rafters is still common. In fact, some carpenters prefer building with rafters. Why? Many carpenters feel that they have more creative control when working with rafters.

A lot of builders prefer to build a rafter roof structure when making a roof for a Cape Cod home that will have dormers. Whether you build with rafters or trusses is up to you. If you choose to use rafters, this chapter will help you to understand maximum rafter spans.

RAFTER SPANS

Rafter spans have limits. We are talking about the unsupported distance that a rafter may extend. The maximum span of rafters varies due to different factors. One factor is the type of wood that the rafter is cut from. Another factor is the type of roof being built. Check your local building code to determine maximum rafter spans in your region. In the meantime, you can check out the many tables that follow in this section to get a good idea of what to expect.

> ▶ *trade* **tip**
>
> When determining the maximum span for rafters, don't stretch the limits. If you have a situation that is borderline on what size rafter to use, choose to use the next larger size. It is far better to have more strength than what is needed, than to have too little.

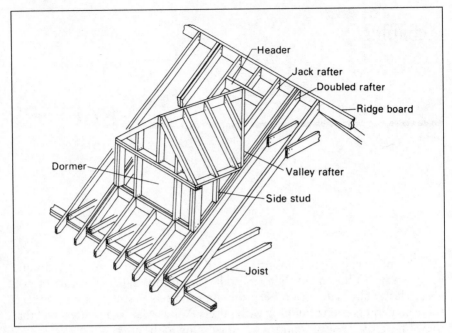

FIGURE 7.1 ■ **Detail of a dormer being built in a rafter roof structure**
(Drawing courtesy of U.S. Dept. of Agriculture)

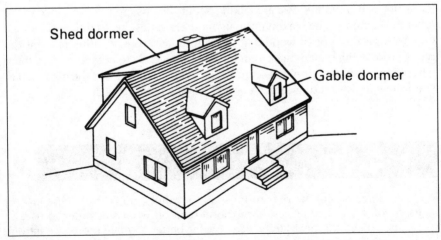

FIGURE 7.2 ■ **Example of a Cape Code home that has both gable dormers and a shed dormer**
(Drawing courtesy of U.S. Dept. of Agriculture)

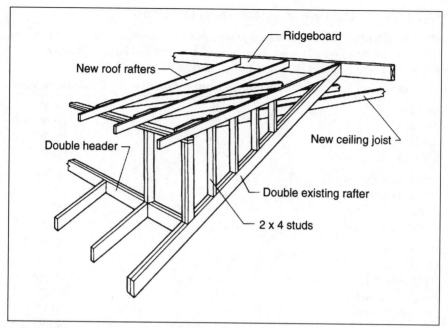

FIGURE 7.3 ■ Framing a shed dormer

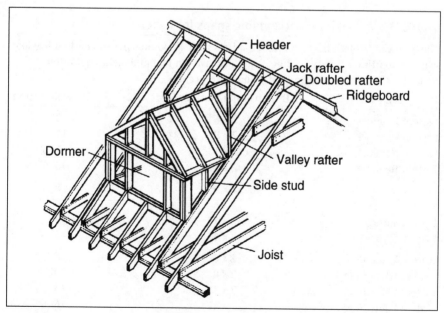

FIGURE 7.4 ■ Framing a gable dormer

TABLE 7.1 ■ Maximum acceptable spans for rafters

The following table is based on sloped rafters with 40 pounds per square foot live load while supporting a drywall ceiling. Spans are given in feet and inches. All rafter material is Grade 2 lumber spaced 16 inches on center.

Species	Dimensional Lumber Size	Allowable Span
Balsam Fir	2 x 6	9' 1"
Balsam Fir	2 x 8	12' 0"
Balsam Fir	2 x 10	15' 3"
Balsam Fir	2 x 12	18' 7"
California Redwood	2 x 6	8' 10"
California Redwood	2 x 8	11' 9"
California Redwood	2 x 10	14' 11"
California Redwood	2 x 12	18' 2"
Douglas Fir-Larch	2 x 6	7' 7"
Douglas Fir-Larch	2 x 8	10' 0"
Douglas Fir-Larch	2 x 10	12' 10"
Douglas Fir-Larch	2 x 12	15' 7"
Douglas Fir-South	2 x 6	9' 1"
Douglas Fir-South	2 x 8	12' 0"
Douglas Fir-South	2 x 10	15' 3"
Douglas Fir-South	2 x 12	18' 7"

TABLE 7.2 ■ Maximum acceptable spans for rafters

The following table is based on sloped rafters with 40 pounds per square foot live load while supporting a drywall ceiling. Spans are given in feet and inches. All rafter material is Grade 2 lumber spaced 16 inches on center.

Species	Dimensional Lumber Size	Allowable Span
Eastern Hemlock	2 x 6	8' 6"
Eastern Hemlock	2 x 8	12' 0"
Eastern Hemlock	2 x 10	15' 3"
Eastern Hemlock	2 x 12	18' 7"
Eastern Spruce	2 x 6	8' 0"
Eastern Spruce	2 x 8	10' 7"
Eastern Spruce	2 x 10	13' 7"
Eastern Spruce	2 x 12	16' 6"
Eastern White Pine	2 x 6	8' 0"
Eastern White Pine	2 x 8	10' 7"
Eastern White Pine	2 x 10	13' 7"
Eastern White Pine	2 x 12	16' 6"
Mountain Hemlock	2 x 6	8' 6"
Mountain Hemlock	2 x 8	11' 4"
Mountain Hemlock	2 x 10	14' 5"
Mountain Hemlock	2 x 12	17' 6"

TABLE 7.3 ▪ Maximum acceptable spans for rafters

The following table is based on sloped rafters with 40 pounds per square foot live load while supporting a drywall ceiling. Spans are given in feet and inches. All rafter material is Grade 2 lumber spaced 16 inches on center.

Species	Dimensional Lumber Size	Allowable Span
Northern Pine	2 x 6	8' 8"
Northern Pine	2 x 8	11' 6"
Northern Pine	2 x 10	14' 8"
Northern Pine	2 x 12	17' 9"
Ponderosa Pine	2 x 6	8' 1"
Ponderosa Pine	2 x 8	10' 9"
Ponderosa Pine	2 x 10	13' 9"
Ponderosa Pine	2 x 12	16' 9"
Southern Pine	2 x 6	9' 9"
Southern Pine	2 x 8	12' 10"
Southern Pine	2 x 10	16' 4"
Southern Pine	2 x 12	19' 11"
Virginia Pine	2 x 6	9' 1"
Virginia Pine	2 x 8	12' 0"
Virginia Pine	2 x 10	15' 3"
Virginia Pine	2 x 12	18' 7"

TABLE 7.4 ▪ Maximum acceptable spans for rafters

The following table is based on sloped rafters with 40 pounds per square foot live load while supporting a drywall ceiling. Spans are given in feet and inches. All rafter material is Grade 2 lumber spaced 24 inches on center.

Species	Dimensional Lumber Size	Allowable Span
Balsam Fir	2 x 6	7' 5"
Balsam Fir	2 x 8	9' 9"
Balsam Fir	2 x 10	12' 6"
Balsam Fir	2 x 12	15' 2"
California Redwood	2 x 6	7' 10"
California Redwood	2 x 8	10' 4"
California Redwood	2 x 10	13' 2"
California Redwood	2 x 12	15' 4"
Douglas Fir-Larch	2 x 6	6' 3"
Douglas Fir-Larch	2 x 8	8' 3"
Douglas Fir-Larch	2 x 10	10' 6"
Douglas Fir-Larch	2 x 12	12' 9"
Douglas Fir-South	2 x 6	7' 10"
Douglas Fir-South	2 x 8	10' 4"
Douglas Fir-South	2 x 10	13' 3"
Douglas Fir-South	2 x 12	16' 1"

TABLE 7.5 ■ Maximum acceptable spans for rafters

The following table is based on sloped rafters with 40 pounds per square foot live load while supporting a drywall ceiling. Spans are given in feet and inches. All rafter material is Grade 2 lumber spaced 24 inches on center.

Species	Dimensional Lumber Size	Allowable Span
Eastern Hemlock	2 x 6	7' 5"
Eastern Hemlock	2 x 8	9' 9"
Eastern Hemlock	2 x 10	12' 6"
Eastern Hemlock	2 x 12	15' 2"
Eastern Spruce	2 x 6	6' 7"
Eastern Spruce	2 x 8	8' 8"
Eastern Spruce	2 x 10	11' 1"
Eastern Spruce	2 x 12	13' 6"
Eastern White Pine	2 x 6	6' 7"
Eastern White Pine	2 x 8	8' 8"
Eastern White Pine	2 x 10	11' 3"
Eastern White Pine	2 x 12	13' 6"
Mountain Hemlock	2 x 6	7' 5"
Mountain Hemlock	2 x 8	9' 9"
Mountain Hemlock	2 x 10	12' 6"
Mountain Hemlock	2 x 12	15' 2"

TABLE 7.6 ■ Maximum acceptable spans for rafters

The following table is based on sloped rafters with 40 pounds per square foot live load while supporting a drywall ceiling. Spans are given in feet and inches. All rafter material is Grade 2 lumber spaced 24 inches on center.

Species	Dimensional Lumber Size	Allowable Span
Northern Pine	2 x 6	7' 1"
Northern Pine	2 x 8	9' 4"
Northern Pine	2 x 10	11' 11"
Northern Pine	2 x 12	14' 6"
Ponderosa Pine	2 x 6	6' 8"
Ponderosa Pine	2 x 8	8' 9"
Ponderosa Pine	2 x 10	11' 3"
Ponderosa Pine	2 x 12	13' 8"
Southern Pine	2 x 6	8' 0"
Southern Pine	2 x 8	10' 7"
Southern Pine	2 x 10	13' 6"
Southern Pine	2 x 12	16' 5"
Virginia Pine	2 x 6	7' 11"
Virginia Pine	2 x 8	10' 5"
Virginia Pine	2 x 10	13' 4"
Virginia Pine	2 x 12	16' 3"

TABLE 7.7 ■ Maximum acceptable spans for rafters

The following table is based on flat or low-sloped rafters with 40 pounds per square foot live load. Slope is based on 3 in 12 or less and no ceiling load. Spans are given in feet and inches. All rafter material is Grade 2 lumber spaced 16 inches on center.

Species	Dimensional Lumber Size	Allowable Span
Balsam Fir	2 x 6	9' 6"
Balsam Fir	2 x 8	12' 7"
Balsam Fir	2 x 10	16' 0"
Balsam Fir	2 x 12	19' 6"
California Redwood	2 x 6	10' 3"
California Redwood	2 x 8	13' 7"
California Redwood	2 x 10	17' 4"
California Redwood	2 x 12	21' 1"
Douglas Fir-Larch	2 x 6	10' 5"
Douglas Fir-Larch	2 x 8	13' 9"
Douglas Fir-Larch	2 x 10	17' 7"
Douglas Fir-Larch	2 x 12	21' 5"
Douglas Fir-South	2 x 6	10' 1"
Douglas Fir-South	2 x 8	13' 4"
Douglas Fir-South	2 x 10	17' 0"
Douglas Fir-South	2 x 12	20' 8"

TABLE 7.8 ■ Maximum acceptable spans for rafters

The following table is based on flat or low-sloped rafters with 40 pounds per square foot live load. Slope is based on 3 in 12 or less and no ceiling load. Spans are given in feet and inches. All rafter material is Grade 2 lumber spaced 16 inches on center.

Species	Dimensional Lumber Size	Allowable Span
Eastern Hemlock	2 x 6	9' 6"
Eastern Hemlock	2 x 8	12' 7"
Eastern Hemlock	2 x 10	16' 0"
Eastern Hemlock	2 x 12	19' 6"
Eastern Spruce	2 x 6	8' 5"
Eastern Spruce	2 x 8	11' 2"
Eastern Spruce	2 x 10	14' 3"
Eastern Spruce	2 x 12	17' 4"
Eastern White Pine	2 x 6	8' 5"
Eastern White Pine	2 x 8	11' 2"
Eastern White Pine	2 x 10	14' 3"
Eastern White Pine	2 x 12	17' 4"
Mountain Hemlock	2 x 6	9' 6"
Mountain Hemlock	2 x 8	12' 7"
Mountain Hemlock	2 x 10	16' 0"
Mountain Hemlock	2 x 12	19' 6"

TABLE 7.9 ■ Maximum acceptable spans for rafters

The following table is based on flat or low-sloped rafters with 40 pounds per square foot live load. Slope is based on 3 in 12 or less and no ceiling load. Spans are given in feet and inches. All rafter material is Grade 2 lumber spaced 16 inches on center.

Species	Dimensional Lumber Size	Allowable Span
Northern Pine	2 x 6	9' 1"
Northern Pine	2 x 8	12' 0"
Northern Pine	2 x 10	15' 4"
Northern Pine	2 x 12	20' 2"
Ponderosa Pine	2 x 6	8' 6"
Ponderosa Pine	2 x 8	11' 4"
Ponderosa Pine	2 x 10	14' 5"
Ponderosa Pine	2 x 12	17' 6"
Southern Pine	2 x 6	10' 3"
Southern Pine	2 x 8	13' 7"
Southern Pine	2 x 10	17' 4"
Southern Pine	2 x 12	21' 1"
Virginia Pine	2 x 6	10' 3"
Virginia Pine	2 x 8	13' 7"
Virginia Pine	2 x 10	17' 4"
Virginia Pine	2 x 12	21' 1"

TABLE 7.10 ■ Maximum acceptable spans for rafters

The following table is based on flat or low-sloped rafters with 40 pounds per square foot live load. Slope is based on 3 in 12 or less and no ceiling load. Spans are given in feet and inches. All rafter material is Grade 2 lumber spaced 24 inches on center.

Species	Dimensional Lumber Size	Allowable Span
Balsam Fir	2 x 6	7' 9"
Balsam Fir	2 x 8	10' 3"
Balsam Fir	2 x 10	13' 1"
Balsam Fir	2 x 12	15' 11"
California Redwood	2 x 6	8' 5"
California Redwood	2 x 8	11' 1"
California Redwood	2 x 10	14' 2"
California Redwood	2 x 12	17' 2"
Douglas Fir-Larch	2 x 6	8' 6"
Douglas Fir-Larch	2 x 8	11' 3"
Douglas Fir-Larch	2 x 10	14' 5"
Douglas Fir-Larch	2 x 12	17' 5"
Douglas Fir-South	2 x 6	8' 3"
Douglas Fir-South	2 x 8	10' 11"
Douglas Fir-South	2 x 10	13' 10"
Douglas Fir-South	2 x 12	16' 10"

TABLE 7.11 ▪ Maximum acceptable spans for rafters

The following table is based on flat or low-sloped rafters with 40 pounds per square foot live load. Slope is based on 3 in 12 or less and no ceiling load. Spans are given in feet and inches. All rafter material is Grade 2 lumber spaced 24 inches on center.

Species	Dimensional Lumber Size	Allowable Span
Eastern Hemlock	2 x 6	7' 9"
Eastern Hemlock	2 x 8	10' 3"
Eastern Hemlock	2 x 10	13' 1"
Eastern Hemlock	2 x 12	15' 11"
Eastern Spruce	2 x 6	6' 11"
Eastern Spruce	2 x 8	9' 1"
Eastern Spruce	2 x 10	11' 7"
Eastern Spruce	2 x 12	14' 1"
Eastern White Pine	2 x 6	6' 11"
Eastern White Pine	2 x 8	9' 1"
Eastern White Pine	2 x 10	11' 7"
Eastern White Pine	2 x 12	14' 1"
Mountain Hemlock	2 x 6	7' 9"
Mountain Hemlock	2 x 8	10' 3"
Mountain Hemlock	2 x 10	13' 1"
Mountain Hemlock	2 x 12	15' 11"

TABLE 7.12 ▪ Maximum acceptable spans for rafters

The following table is based on flat or low-sloped rafters with 40 pounds per square foot live load. Slope is based on 3 in 12 or less and no ceiling load. Spans are given in feet and inches. All rafter material is Grade 2 lumber spaced 24 inches on center.

Species	Dimensional Lumber Size	Allowable Span
Northern Pine	2 x 6	7' 5"
Northern Pine	2 x 8	9' 10"
Northern Pine	2 x 10	12' 6"
Northern Pine	2 x 12	15' 3"
Ponderosa Pine	2 x 6	7' 0"
Ponderosa Pine	2 x 8	9' 2"
Ponderosa Pine	2 x 10	11' 9"
Ponderosa Pine	2 x 12	14' 3"
Southern Pine	2 x 6	8' 5"
Southern Pine	2 x 8	11' 1"
Southern Pine	2 x 10	14' 2"
Southern Pine	2 x 12	17' 2"
Virginia Pine	2 x 6	8' 5"
Virginia Pine	2 x 8	11' 1"
Virginia Pine	2 x 10	14' 2"
Virginia Pine	2 x 12	17' 2"

TABLE 7.13 ■ Maximum acceptable spans for rafters

The following table is based on medium or high sloped rafters with 40 pounds per square foot live load. Slope is based on 3 in 12 or less and no ceiling load. Spans are given in feet and inches. All rafter material is Grade 2 lumber spaced 16 inches on center.

Species	Dimensional Lumber Size	Allowable Span
Balsam Fir	2 x 4	5' 9"
Balsam Fir	2 x 6	9' 1"
Balsam Fir	2 x 8	12' 0"
Balsam Fir	2 x 10	15' 3"
California Redwood	2 x 4	6' 3"
California Redwood	2 x 6	9' 10"
California Redwood	2 x 8	12' 11"
California Redwood	2 x 10	16' 6"
Douglas Fir-Larch	2 x 4	6' 11"
Douglas Fir-Larch	2 x 6	10' 10"
Douglas Fir-Larch	2 x 8	14' 3"
Douglas Fir-Larch	2 x 10	18' 2"
Douglas Fir-South	2 x 4	6' 1"
Douglas Fir-South	2 x 6	9' 7"
Douglas Fir-South	2 x 8	12' 8"
Douglas Fir-South	2 x 10	16' 2"

TABLE 7.14 ■ Maximum acceptable spans for rafters

The following table is based on medium or high sloped rafters with 40 pounds per square foot live load. Slope is based on 3 in 12 or less and no ceiling load. Spans are given in feet and inches. All rafter material is Grade 2 lumber spaced 16 inches on center.

Species	Dimensional Lumber Size	Allowable Span
Eastern Hemlock	2 x 4	5' 9"
Eastern Hemlock	2 x 6	9' 1"
Eastern Hemlock	2 x 8	12' 0"
Eastern Hemlock	2 x 10	15' 3"
Eastern Spruce	2 x 4	5' 1"
Eastern Spruce	2 x 6	8' 0"
Eastern Spruce	2 x 8	10' 7"
Eastern Spruce	2 x 10	13' 7"
Eastern White Pine	2 x 4	5' 1"
Eastern White Pine	2 x 6	8' 0"
Eastern White Pine	2 x 8	10' 7"
Eastern White Pine	2 x 10	13' 7"
Mountain Hemlock	2 x 4	5' 9"
Mountain Hemlock	2 x 6	9' 1"
Mountain Hemlock	2 x 8	12' 0"
Mountain Hemlock	2 x 10	15' 3"

TABLE 7.15 ■ Maximum acceptable spans for rafters

The following table is based on medium or high sloped rafters with 40 pounds per square foot live load. Slope is based on 3 in 12 or less and no ceiling load. Spans are given in feet and inches. All rafter material is Grade 2 lumber spaced 16 inches on center.

Species	Dimensional Lumber Size	Allowable Span
Northern Pine	2 x 4	5' 6"
Northern Pine	2 x 6	8' 8"
Northern Pine	2 x 8	11' 6"
Northern Pine	2 x 10	14' 8"
Ponderosa Pine	2 x 4	5' 2"
Ponderosa Pine	2 x 6	8' 1"
Ponderosa Pine	2 x 8	10' 9"
Ponderosa Pine	2 x 10	13' 9"
Southern Pine	2 x 4	6' 3"
Southern Pine	2 x 6	9' 10"
Southern Pine	2 x 8	12' 11"
Southern Pine	2 x 10	16' 6"
Virginia Pine	2 x 4	6' 3"
Virginia Pine	2 x 6	9' 10"
Virginia Pine	2 x 8	12' 11"
Virginia Pine	2 x 10	16' 6"

TABLE 7.16 ■ Maximum acceptable spans for rafters

The following table is based on medium or high sloped rafters with 40 pounds per square foot live load. Slope is based on 3 in 12 or less and no ceiling load. Spans are given in feet and inches. All rafter material is Grade 2 lumber spaced 24 inches on center.

Species	Dimensional Lumber Size	Allowable Span
Balsam Fir	2 x 4	4' 9"
Balsam Fir	2 x 6	7' 5"
Balsam Fir	2 x 8	9' 9"
Balsam Fir	2 x 10	12' 6"
California Redwood	2 x 4	5' 1"
California Redwood	2 x 6	8' 0"
California Redwood	2 x 8	10' 7"
California Redwood	2 x 10	13' 6"
Douglas Fir-Larch	2 x 4	5' 2"
Douglas Fir-Larch	2 x 6	8' 1"
Douglas Fir-Larch	2 x 8	10' 9"
Douglas Fir-Larch	2 x 10	13' 8"
Douglas Fir-South	2 x 4	5' 0"
Douglas Fir-South	2 x 6	7' 10"
Douglas Fir-South	2 x 8	10' 4"
Douglas Fir-South	2 x 10	13' 3"

TABLE 7.17 ■ Maximum acceptable spans for rafters

The following table is based on medium or high sloped rafters with 40 pounds per square foot live load. Slope is based on 3 in 12 or less and no ceiling load. Spans are given in feet and inches. All rafter material is Grade 2 lumber spaced 24 inches on center.

Species	Dimensional Lumber Size	Allowable Span
Eastern Hemlock	2 x 4	4' 9"
Eastern Hemlock	2 x 6	7' 5"
Eastern Hemlock	2 x 8	9' 9"
Eastern Hemlock	2 x 10	12' 6"
Eastern Spruce	2 x 4	4' 2"
Eastern Spruce	2 x 6	6' 7"
Eastern Spruce	2 x 8	8' 8"
Eastern Spruce	2 x 10	11' 1"
Eastern White Pine	2 x 4	4' 2"
Eastern White Pine	2 x 6	6' 7"
Eastern White Pine	2 x 8	8' 8"
Eastern White Pine	2 x 10	11' 1"
Mountain Hemlock	2 x 4	4' 9"
Mountain Hemlock	2 x 6	7' 5"
Mountain Hemlock	2 x 8	9' 9"
Mountain Hemlock	2 x 10	12' 6"

TABLE 7.18 ■ Maximum acceptable spans for rafters

The following table is based on medium or high sloped rafters with 40 pounds per square foot live load. Slope is based on 3 in 12 or less and no ceiling load. Spans are given in feet and inches. All rafter material is Grade 2 lumber spaced 24 inches on center.

Species	Dimensional Lumber Size	Allowable Span
Northern Pine	2 x 4	4' 6"
Northern Pine	2 x 6	7' 1"
Northern Pine	2 x 8	9' 4"
Northern Pine	2 x 10	11' 11"
Ponderosa Pine	2 x 4	4' 3"
Ponderosa Pine	2 x 6	6' 8"
Ponderosa Pine	2 x 8	8' 9"
Ponderosa Pine	2 x 10	11' 3"
Southern Pine	2 x 4	5' 1"
Southern Pine	2 x 6	8' 0"
Southern Pine	2 x 8	10' 7"
Southern Pine	2 x 10	13' 6"
Virginia Pine	2 x 4	5' 1"
Virginia Pine	2 x 6	8' 0"
Virginia Pine	2 x 8	10' 7"
Virginia Pine	2 x 10	13' 6"

☑ *fast*facts

The theoretical length of a common rafter is the shortest distance between the outer edge of a plate and a point where the measuring line of a rafter comes into contact with a ridge line.

RAFTER FRAMING

Rafter framing requires one end of a rafter to rest on a ridge board while the other end of the rafter rests on a top plate. Getting the first rafter cut properly is the key to success. Once one rafter is cut to the proper angles, that rafter can be used as a pattern to cut remaining rafters.

The ends of rafters can be cut in different ways. A bird's mouth is cut to allow the rafter to sit properly on a top plate. Then the rafter tail is cut in a manner that will provide the type of aesthetics desired.

Once the cuts are made and a rafter is in place, it is nailed to the top plate and the ridge board. Most framing contractors today use pneumatic nailers. Since the nailing of rafters is done with the toenail method, you may have to adjust your nail gun to achieve the proper depth setting of nails during the framing process.

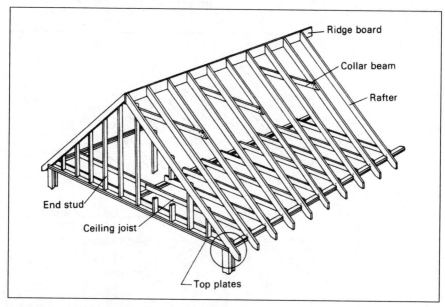

FIGURE 7.5 ■ **Components of rafter framing**
(Drawing courtesy of U.S. Dept. of Agriculture)

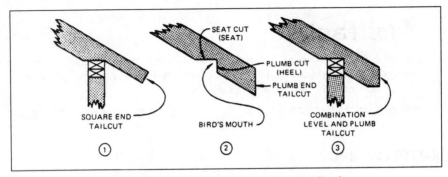

FIGURE 7.6 ■ Common cuts on bottoms and ends of rafters

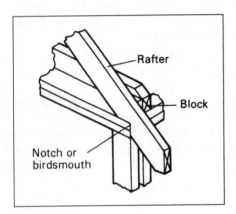

FIGURE 7.7 ■ Example of a bird's mouth cut into a rafter

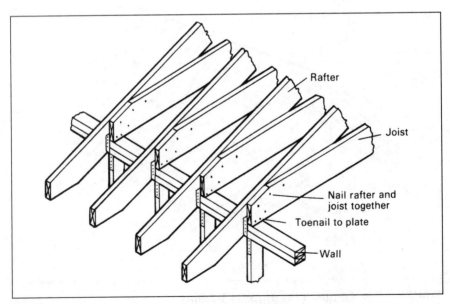

FIGURE 7.8 ■ Toenailing rafters to a top plate
(Drawing courtesy of U.S. Dept. of Agriculture)

Safety is a major issue when framing a roof structure. Your personal safety is one concern, but always be aware of what is below you. A rafter that falls from the work area can be extremely dangerous. Any tool dropped can injure someone below. Make sure that your jobsite safety plan is in place during all work.

ROOF PITCH

Calculating the pitch of a roof is not difficult. There are only two measurements needed to do the math. First, measure from the top surface of the top plate to the peak of the roof on a gable end. Mark the centerline of the horizontal span where the roof rises to its peak. Measure the distance from an outside top plate to the centerline mark on the horizontal run. Let's assume that you are dealing with a foundation that is 24 feet wide and that the roof peak is centered over the foundation. This makes your horizontal measurement 12 feet. Further assume that the distance from the top plate to the peak of the roof is 8 feet. This gives you a roof pitch of 8-12. If the vertical rise had been 6 feet, the roof pitch would be 6-12. A vertical rise of 10 feet would give a 10-12 pitch. The larger the first number is in the pitch, the steeper the roof is.

The run of a roof is generally equal to half the total width of a building.

A rafter square can be a carpenter's best friend when it comes to building a roof structure. The square has exhaustive data on it that will allow you to tell the length of any hip or valley rafter per foot of run. You can also determine the line length of any pitch with a square. Side cuts for jack, hip, and valley rafters can be calculated with a square. Get a good rafter square and learn how to use it.

☑ *fast***facts**

Specialized calculators are available for the building trade. These devices can prove to be very valuable. Some carpenters prefer their framing square over a calculator, but having a trade-specific calculator can be of great benefit to you.

HOW DO I USE A RAFTER SQUARE?

How do I use a framing square? When you look at a rafter square, you will find a lot of numbers in various sections. The method of using a square depends on the job at hand. As an example, let's explore how you would determine the length of a common rafter for a roof that has an 8-12 pitch. For the sake of this example, assume that the span of the roof is 34 feet. Using the rule-of-thumb that the run is half the span, our run is 17 feet. With this known, you can refer to your rafter square to find the number that represents the per-foot-of-run rafter number. Where can this be found? It will be right below the large numbers on the square. In our case, look directly below the number 8. You will find the number 14.42. This represents, in inches, the rafter length per foot of run. The number above it, the 8, represents the rise, in inches, for a roof. To summarize to this point, we have a roof with an 8-12 pitch, a run of 17 feet, and a rafter length per foot of run established at 14.42

The next step is to multiply the length per foot of run by the number of the feet of run. This means that we multiply 14.42 inches by 17. This gives us 245.14 inches. To get this measurement converted to feet, we divide by 12. Our answer will be 20.42 feet. In round numbers, we have 20.5 feet. This is the length of our rafter, except that we have to add length for any overhang. If we assume a 12-inch overhang, our finished rafter length will be 21.5 feet.

When you are figuring and cutting rafters, you should not be without a good rafter square. Small Speed™ squares are also very helpful in doing framing math. My suggestion to you is to get both a quality rafter square and a quality Speed™ square. Once you have them, spend time learning how to use them. The time invested in the learning curve will be returned many times over on job sites.

Once you have a good square and the knowledge of how to use it, you can design and lay out most any type of roof structure. You could spend a lot of time reading math books and doing all sorts of geometry and other math, or you can do what experienced carpenters do and use a rafter square for your math requirements.

chapter 8

ENGINEERED TRUSSES

Engineered trusses are the modern equivalent of rafters. Trusses are, in fact, more versatile than rafters. It is amazing how much load trusses can hold when they are engineered for a specific job. The span for trusses is much longer than it would be for a rafter system.

Trusses can be built with smaller lumber than what would be needed for rafters holding the same load. Since trusses are usually built off site and trucked in, the cost of trusses can be very attractive to builders. While rafters offer certain advantages, trusses are common in modern construction.

Some builders prefer to make their own trusses. This practice is rare. When all time and expense is considered, buying trusses made in a factory usually makes the most sense. Most builders have their trusses trucked to the job site and often have them set in place with the use of a crane. This allows a framing contractor to create a roof structure much faster than building it out of rafters.

Most trusses are triangular in shape, but an engineered truss can take on different shapes. All trusses are engineered for their specific use. This allows the truss to handle a maximum load at a maximum span. Even 2 x 4 lumber can be used to build trusses. That's right, I said 2 x 4 lumber.

☑ fast**facts**

Trusses must be handled with care. Rough treatment can damage a truss. If a truss is damaged, it cannot be used without competent repair by a suitable professional. Damaged trusses can be repaired, but the repair must meet specific engineering requirements.

TRUSS CONFIGURATIONS

TRUSS TERMS: The terms below are typically used to describe the various parts of a metal plate connected wood truss. The truss profile, span, heel height, overall height, overhang and web configuration depend on the specific design conditions and will vary by application.

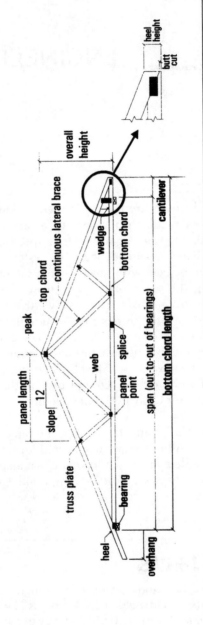

FIGURE 8.1 ■ Components of a roof truss

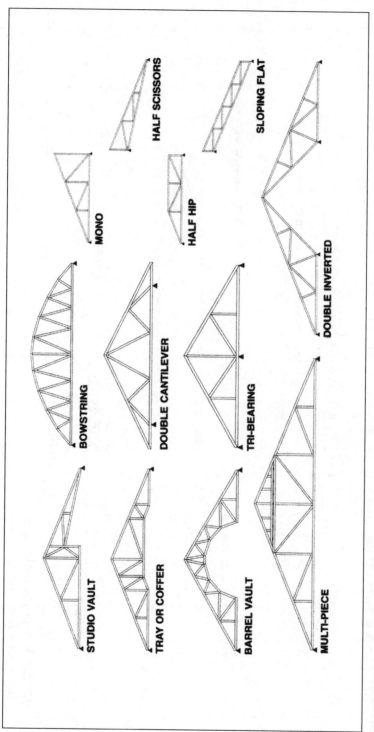

FIGURE 8.2 ■ Types of roof trusses
(Courtesy of Wood Truss Council of America)

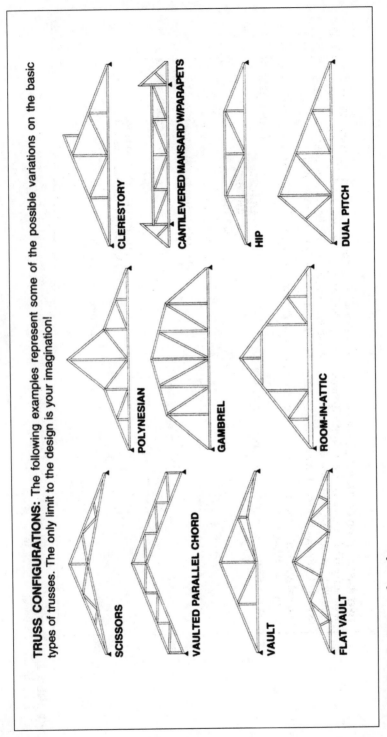

TRUSS CONFIGURATIONS: The following examples represent some of the possible variations on the basic types of trusses. The only limit to the design is your imagination!

SCISSORS

VAULTED PARALLEL CHORD

VAULT

FLAT VAULT

POLYNESIAN

GAMBREL

ROOM-IN-ATTIC

CLERESTORY

CANTILEVERED MANSARD W/PARAPETS

HIP

DUAL PITCH

FIGURE 8.3 ■ **Types of roof trusses**
(Courtesy of Wood Truss Council of America)

TWO BASIC TYPES OF TRUSSES: The pitched or common truss is characterized by its triangular shape. It is most often used for roof construction. Some common trusses are named according to their web configuration, such as the King Post, Fan, Fink or Howe truss. The chord size and web configuration are determined by span, load, and spacing. All truss designs are optimized to provide the most economical application.

The parallel chord or flat truss gets its name from having parallel top and bottom chords. This type is often used for floor construction.

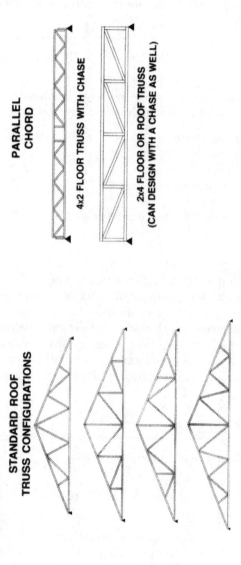

STANDARD ROOF
TRUSS CONFIGURATIONS

PARALLEL
CHORD

4x2 FLOOR TRUSS WITH CHASE

2x4 FLOOR OR ROOF TRUSS
(CAN DESIGN WITH A CHASE AS WELL)

FIGURE 8.4 ■ Standard and parallel roof and floor trusses
(Courtesy of Wood Truss Council of America)

▶ *trade* **tip**

Engineered trusses must not be cut or altered. Since the trusses are engineered to tight standards, any alteration can affect the structural integrity of the truss.

What types of trusses are available beyond the typical triangle truss? There are many, some of which include the following:

- Bowstring trusses
- Mono trusses
- Half hip trusses
- Sloping flat trusses
- Hip trusses
- Cantilevered Mansard trusses

SPEED

The speed with which a roof structure can be applied to a building when using trusses is a major reason for the popularity of engineered trusses. But, this is not the only reason trusses are used. For example, some interior partitions that would be needed with a rafter system can be eliminated with a truss system. This is due to the fact that trusses can span greater distances without support. This can become especially important if you are building a storage building, barn, or similar structure. If you need open floor space, trusses are the way to go.

TOP THREE TRUSSES

The top three types of trusses are:

- King-post trusses
- Fink trusses, also known as W-trusses
- Scissor trusses

The king-post truss is the simplest type of truss that is usually used in home construction. King-post trusses are more limited in their span length than Fink trusses are. When short to medium spans are needed, the king-post trust is economical.

Fink trusses are widely used and very popular. Then there are scissor trusses. These are used for buildings with sloping ceilings. Most homebuilders will be satisfied with one of the three top trusses, but there are many other types to choose from for special applications.

▶ *trade* **tip**

Trusses are frequently engineered to be spaced on 24-inch centers. Due to this spacing, it can be necessary to install thicker interior and exterior sheathing.

TRUSS SPANS

Truss spans vary according to conditions. The following tables will aid you in determining suitable spans:

While trusses are likely to continue growing in popularity, there will probably always be a place in construction for rafters. Even if you prefer rafters over trusses, it may help your bottom line to investigate using trusses. They are definitely worth a try.

TABLE 8.1 ▪ Truss spans

Chord size: 2-×-4 top and 2-×-4 bottom	
Pitch	Span (feet)
2/12	22
3/12	29
4/12	33
5/12	35
6/12	37

Typical truss spans (55 PSF with 15% duration factor).

Chord size: 2-×-6 top and 2-×-6 bottom	
Pitch	Span (feet)
2/12	32
3/12	51
4/12	56
5/12	60
6/12	62

Monopitch truss spans (55 PSF with 33% duration factor).

Chord size: 2-×-6 top and 2-×-4 bottom	
Pitch	Span (feet)
2/12	28
3/12	39
4/12	46
5/12	53
6/12	57

Typical truss spans (47 PSF with 33% duration factor).

TABLE 8.2 ■ Truss spans

Chord size: 2-x-4 top and 2-x-4 bottom	
Pitch	Span (feet)
2/12	25
3/12	33
4/12	37
5/12	40
6/12	41

Monopitch truss spans (55 PSF with 33% duration factor).

Chord size: 2-x-6 top and 2-x-4 bottom	
Pitch	Span (feet)
2/12	23
3/12	31
4/12	39
5/12	45
6/12	51

Monopitch truss spans (55 PSF with 15% duration factor).

Chord size: 2-x-4 top and 2-x-4 bottom	
Pitch	Span (feet)
2/12	22
3/12	30
4/12	33
5/12	35
6/12	37

Monopitch truss spans (55 PSF with 15% duration factor).

TABLE 8.3 ■ Truss spans

Chord size: 2-x-6 top and 2-x-6 bottom	
Pitch	Span (feet)
2/12	34
3/12	45
4/12	50
5/12	53
6/12	56

Monopitch truss spans (55 PSF with 15% duration factor).

Chord size: 2-x-6 top and 2-x-4 bottom	
Pitch	Span (feet)
2/12	25
3/12	35
4/12	42
5/12	48
6/12	53

Monopitch truss spans (55 PSF with 33% duration factor).

Chord size: 2-x-6 top and 2-x-6 bottom	
Pitch	Span (feet)
2/12	44
3/12	57
4/12	63
5/12	67
6/12	68

Monopitch truss spans (47 PSF with 33% duration factor).

TABLE 8.4 ■ Truss spans

Chord size: 2-×-4 top and 2-×-4 bottom	
Pitch	Span (feet)
2/12	28
3/12	38
4/12	42
5/12	44
6/12	45

Monopitch truss spans (47 PSF with 33% duration factor).

Chord size: 2-×-6 top and 2-×-4 bottom	
Pitch	Span (feet)
2/12	28
3/12	38
4/12	46
5/12	52
6/12	57

Typical truss spans (47 PSF with 33% duration factor).

Chord size: 2-×-4 top and 2-×-4 bottom	
Pitch	Span (feet)
2/12	28
3/12	37
4/12	41
5/12	44
6/12	44

Typical truss spans (47 PSF with 33% duration factor).

TABLE 8.5 ■ Truss spans

Chord size: 2-×-6 top and 2-×-4 bottom	
Pitch	Span (feet)
2/12	25
3/12	34
4/12	42
5/12	48
6/12	53

Typical truss spans (55 PSF with 33% duration factor).

Chord size: 2-×-4 top and 2-×-4 bottom	
Pitch	Span (feet)
2/12	25
3/12	33
4/12	37
5/12	39
6/12	41

Typical truss spans (55 PSF with 33% duration factor).

Chord size: 2-×-6 top and 2-×-6 bottom	
Pitch	Span (feet)
2/12	39
3/12	50
4/12	55
5/12	59
6/12	62

Typical truss spans (55 PSF with 33% duration factor).

TABLE 8.6 ■ Truss spans

Chord size: 2-x-6 top and 2-x-4 bottom	
Pitch	**Span (feet)**
2/12	23
3/12	31
4/12	39
5/12	45
6/12	51

Typical truss spans (55 PSF with 15% duration factor).

Chord size: 2-x-6 top and 2-x-6 bottom	
Pitch	**Span (feet)**
2/12	34
3/12	44
4/12	49
5/12	53
6/12	55

Typical truss spans (55 PSF with 15% duration factor).

Chord size: 2-x-4 top and 2-x-4 bottom		
Top chord pitch	**Bottom chord pitch**	**Span (feet)**
6/12	2/12	36
6/12	3/12	31
6/12	4/12	24

Scissor truss spans (55 PSF with 33% duration factor).

TABLE 8.7 ■ Truss spans

Chord size: 2-x-6 top and 2-x-6 bottom		
Top chord pitch	**Bottom chord pitch**	**Span (feet)**
6/12	2/12	54
6/12	3/12	48
6/12	4/12	36

Scissor truss spans (55 PSF with 33% duration factor).

Chord size: 2-x-6 top and 2-x-4 bottom		
Top chord pitch	**Bottom chord pitch**	**Span (feet)**
6/12	2/12	38
6/12	3/12	30
6/12	4/12	22

Scissor truss spans (55 PSF with 15% duration factor).

Chord size: 2-x-4 top and 2-x-4 bottom		
Top chord pitch	**Bottom chord pitch**	**Span (feet)**
6/12	2/12	32
6/12	3/12	28
6/12	4/12	21

Scissor truss spans (55 PSF with 15% duration factor).

Chord size: 2-x-6 top and 2-x-6 bottom		
Top chord pitch	**Bottom chord pitch**	**Span (feet)**
6/12	2/12	48
6/12	3/12	42
6/12	4/12	32

Scissor truss spans (55 PSF with 15% duration factor).

TABLE 8.8 ■ Truss spans

| Chord size: 2-×-6 top and 2-×-4 bottom | | |
Top chord pitch	Bottom chord pitch	Span (feet)
6/12	2/12	42
6/12	3/12	34
6/12	4/12	24

Scissor truss spans (55 PSF with 33% duration factor).

| Chord size: 2-×-6 top and 2-×-6 bottom | | |
Top chord pitch	Bottom chord pitch	Span (feet)
6/12	2/12	61
6/12	3/12	54
6/12	4/12	42

Scissor truss spans (47 PSF with 33% duration factor).

| Chord size: 2-×-6 top and 2-×-4 bottom | | |
Top chord pitch	Bottom chord pitch	Span (feet)
6/12	2/12	46
6/12	3/12	38
6/12	4/12	27

Scissor truss spans (47 PSF with 33% duration factor).

| Chord size: 2-×-4 top and 2-×-4 bottom | | |
Top chord pitch	Bottom chord pitch	Span (feet)
6/12	2/12	40
6/12	3/12	35
6/12	4/12	27

Scissor truss spans (47 PSF with 33% duration factor).

chapter 9

ROOF SHEATHING
AND COVERINGS

Figuring roof sheathing and coverings can be pretty simple. A gable roof makes doing the math easy. But, not all roof designs are so easy. What makes a gable so easy to compute? Well, if you know the length of the building and the distance from the edge of the roof to the peak, the math is very simple. A Gambrel design is a bit more difficult to figure, but it still is not bad. Don't worry, you are going to see how easy calculating roof requirements can be.

Roofing requires sheathing. This sheathing is usually applied in 4' x 8' sheets. However, some framers still use boards to create a roofing platform. Due to market conditions, 4' x 8' sheets of sheathing can be difficult to obtain. This is especially true during hurricane seasons, wars, and other events that require boarding up buildings. This is usually when builders revert back to board roofs.

FIGURING SHEATHING

Figuring sheathing for a roof is not difficult. The math is done based on the number of square feet of coverage needed. A 4' x 8' sheet of plywood will cover 32 square feet. If you are building with boards, you have to calculate the needs based on square feet or board feet. For example, if you are using a board that is 12 inches wide and 10 feet long, that board contains 10 square feet. This type of math will not make many people scratch their head.

▶ *trade* **tip**

When using plywood, or similar sheathing, that does not have its edge on a rafter or truss, use plywood clips to support the edges of the sheathing.

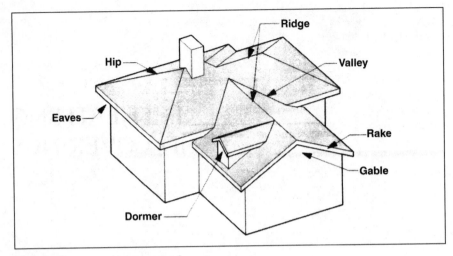

FIGURE 9.1 ■ **Common roof parts**

Roof estimates, whether for sheathing or shingles, are done in terms of square feet. For example, one square of shingles will cover 100 square feet. In the case of 3-tab asphalt shingles, there are three bundles of shingles per square. So how much sheathing will you need for a gable roof that is 40 feet in length with a measurement from the drip edge to the ridge of 16 feet?

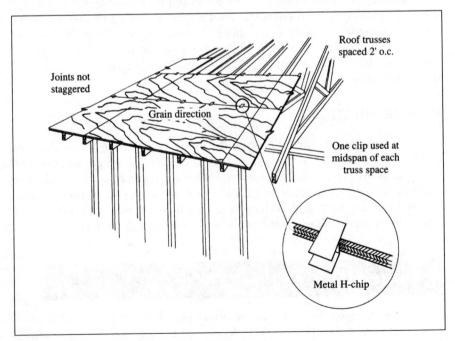

FIGURE 9.2 ■ **Plywood clip installation**

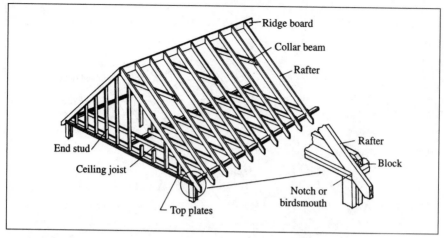

FIGURE 9.3 ■ Typical rafter framing as you might experience when computing roof sheathing

To determine the amount of sheathing needed, the first step is to convert the roof area to square feet. There are two sides of the roof to take into consideration. To do this math, your equation should look like this:

$$40' \times 16' = 640 \times 2 = 1280 \text{ square feet}$$

The procedure above gives you the square footage for one side of the roof and is then multiplied by two in order to determine the total square footage. Now let's assume that you are going to use 4' x 8' sheets of plywood for sheathing. You know that each sheet contains 32 square feet, so how many sheets do you need? Do the math by dividing the total square footage, in this case 1280, by 32. The correct answer will be 40 sheets. Most contractors fac-

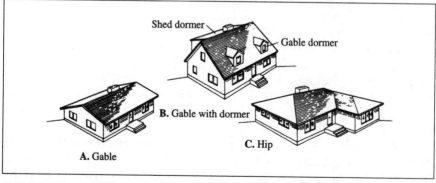

FIGURE 9.4 ■ Different roof designs will require different mathematical approaches to calculate roofing needs.

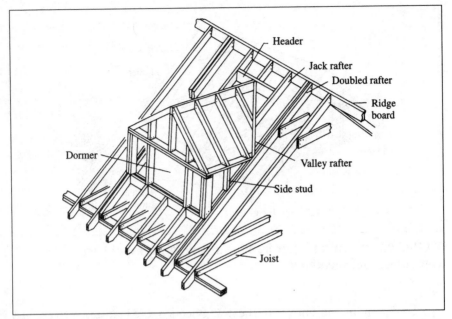

FIGURE 9.5 ■ Gable roof with a dormer

tor in some percentage of the total for wasted materials. It is not unusual for a contractor to add 15 percent for waste.

You will have to use different types of math to determine the roofing needs for various types of roofs. We will cover this type of math shortly. One thing to remember is something that may seem troublesome probably isn't. For example, a gable roof with dormers can be figured with the same method that you have just learned. You simply factor in the amount of roofing needed for the dormer roof and subtract the surface area of the main roof that is occupied by the dormer.

HIP ROOFS

Hip roofs intimidate some people when it comes to figuring the square footage of a roof. The math really is not difficult. Assume that you have a hip roof on a square building. This type of roof is easy to figure. The hip roof is made up of four triangles. Most carpenters know how to calculate gable ends for siding, and the same practice can be used here for estimating the roofing requirements.

Since you have four equal triangles, all you have to do is determine the area of one triangle and multiply it by four. This requires you to measure the base of the triangle. Assume that it is 20 feet wide. Next you measure the distance from the base of the triangle to the point. Assume that this results in a measurement of 15 feet. Now multiply the two numbers and you will get an

☑ *fast***facts**

A shed roof is the easiest type of roof to calculate the square footage. You simply multiply the length by the width to determine the total area.

answer of 300 square feet. This would be accurate if you were figuring a square, but you are figuring a triangle, so you have half the area, or 150 square feet. Once you have this number, multiply it by four and get an answer of 600 square feet. This is very simple math, but it works.

If you encounter a hip roof on a rectangular building, the math is more difficult. The first step for this type of roof is to multiply the length of the eave edge along the short side of the rectangle by the height of the triangle that is formed by the roof on that side. The next step requires you to determine the length of the eave edge along the long side of the rectangle. You should add this figure to the length of the ridge line and multiply the sum by the height of the line measure from eave edge to ridge line. Then you add the results of the first two steps to obtain the total roof area. This sounds complicated and can be difficult to follow at first, but once you do it in the field, the procedure will become much easier.

GAMBREL ROOFS

Gambrel roofs are not difficult to figure. If you think of each segment of the gambrel as a shed roof, the procedure becomes simple. A gambrel roof has four sections. Take the width of each section and multiply it by the length of that section to determine the square footage.

SLOPE MULTIPLIERS

Slope multipliers can be used to determine roof area. This is done by finding the ground area of a building and using an established multiplier to arrive at a correct answer. Assume that you have a house with a foundation size of 24 feet by 40 feet. The house has a gable roof with a 6/12 pitch. Refer to Table 9.1 to determine your multiplier. The correct multiplier is 1.22 Now that you have this information, you can determine the roof area with the following formula:

$$24 \times 40 \times 1.22 = 1171.2 \text{ square feet}$$

▶ *trade* **tip**

Slope multipliers can be used to convert ground area to roof area.

TABLE 9.1 ■ Slope multipliers

Roof Slope	Multiplier
2/12	1.11
3/12	1.13
4/12	1.15
5/12	1.18
6/12	1.22
8/12	1.30
10/12	1.40
12/12	1.51

ASPHALT SHINGLES

Asphalt shingles are the most common type of roofing found on residential properties. Within this type of roofing there can be many variations. Not all types of asphalt shingles provide the same amount of coverage. Pay attention to the slope of a roof when using slope multipliers. Roofing materials are heavy, don't move more than you need on to a roof.

Before you install asphalt shingles, install 15-pound felt on the roof structure. Failure to install the felt may result in a voided warranty on the shingles. How much 15-pound felt will you need for a roof that has 1200 square feet of area? A standard roll of 15-pound felt will produce about 400 square feet of coverage, so you would need three rolls for a roof with 1200 square feet of area.

A common mistake made when installing roofing is the nailing of the shingles. Roofers sometimes get sloppy. Pay attention to your nailing patterns and procedures to avoid angry customers.

TABLE 9.2 ▪ Defining roof slope as a percentage and inches per foot

Defining Roof Slopes and other Types of Slopes

Percent Slope	Inch/Ft	Ratio	Degrees from Horizontal
1%	1/8	1 in 100	—
2%	1/4	1 in 50	—
3%	3/8	—	—
4%	1/2	1 in 25	—
5%	5/8	1 in 20	3
6%	3/4	—	—
7%	7/8	—	—
8%	approx. 1	approx. 1 in 12	—
9%	1 1/8	—	—
10%	1 1/4	1 in 10	6
11%	1 3/8	approx. 1 in 9	—
12%	1 1/2	—	—
13%	1 5/8	—	—
14%	1 3/4	—	—
15%			8.5
16%	1 7/8	—	—
17%	2	approx. 2 in 12	—
18%	2 1/8	—	—
19%	2 1/4	—	—
20%	2 3/8	1 in 5	11.5
25%	3	3 in 12	14
30%	3.6	1 in 3.3	17
35%	4.2	approx. 4 in 12	19.25
40%	4.8	approx. 5 in 12	21.5
45%	5.4	1 in 22	24
50%	6	6 in 12	26.5
55%	6 5/8	1 in 1.8	28.5
60%	7 1/4	approx. 7 in 12	31
65%	7 3/4	1 in 1 1/2	33
70%	8 1/8	1 in 1.4	35
75%	9	1 in 1.3	36.75
100%	12	1 in 1	45

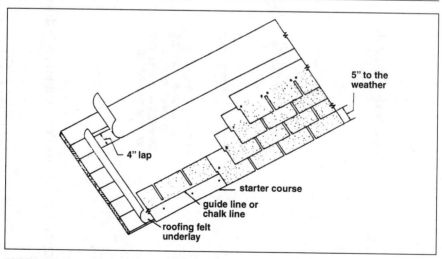

FIGURE 9.6 ▪ Cutaway of typical roofing on a board roof

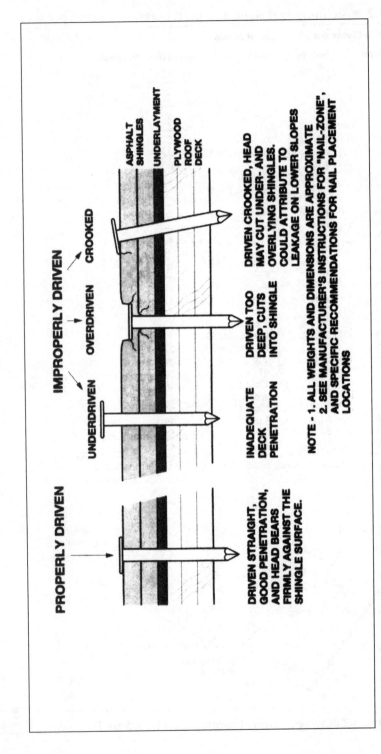

FIGURE 9.7 ■ Properly and improperly driven nails

TABLE 9.3 ▪ Slate shingle size, weight, and exposures

Slate Thickness	Sloping Roof With 3" (76mm) Lap (approx. Pounds Per Square [kg/m²])
³⁄₁₆" to ¼" (4mm to 6mm)	700 to 1,000 lbs/sq (3,417 to 4,882 kg/m²)
⅜" (9mm)	1,500 lbs/sq (7,323 kg/m²)
½" (13mm)	2,000 lbs/sq (9,764 kg/m²)
¾" (19mm)	3,000 lbs/sq (14,646 kg/m²)
1" (25mm)	4,000 lbs/sq (19,528 kg/m²)
1¼" (32mm)	5,000 lbs/sq (24,410 kg/m²)
1½" (38mm)	6,000 lbs/sq (29,292 kg/m²)
1¾" (44mm)	7,000 lbs/sq (34,174 kg/m²)
2" (51mm)	8,000 lbs/sq (39,056 kg/m²)

OTHER TYPES OF ROOFING

There are other types of roofing materials to consider for certain jobs. You might use slate in special circumstances. Cedar shakes are popular in some areas. As I mentioned earlier, check the manufacturer's recommendations for coverage requirements for any type of roofing that you choose to use.

TABLE 9.4 ▪ Schedule for standard slate

SCHEDULE FOR STANDARD 3/16"(5mm) THICK SLATE

SIZE OF SLATE (L x W) (IN.)	SIZE OF SLATE (L x W) (MM)	SLATES PER SQUARE	EXPOSURE WITH 3"(76mm) Lap (IN.)	EXPOSURE WITH 3"(76mm) Lap (MM)	SIZE OF SLATE (L x W) (IN.)	SIZE OF SLATE (L x W) (MM)	SLATES PER SQUARE	EXPOSURE WITH 3"(76mm) Lap (IN.)	EXPOSURE WITH 3"(76mm) Lap (MM)
26 x 14	660 x 356	89	11 1/2	292	16 x 14	406 x 356	160	6 1/2	165
					16 x 12	406 x 305	184	6 1/2	165
24 x 16	610 x 406	86	10 1/2	267	16 x 11	406 x 279	201	6 1/2	165
24 x 14	610 x 356	98	10 1/2	267	16 x 10	406 x 254	222	6 1/2	165
24 x 13	610 x 330	106	10 1/2	267	16 x 9	406 x 229	246	6 1/2	165
24 x 12	610 x 305	114	10 1/2	267	16 x 8	406 x 203	277	6 1/2	165
24 x 11	610 x 279	138	10 1/2	267					
22 x 14	559 x 356	108	9 1/2	241	14 x 12	356 x 305	218	5 1/2	140
22 x 13	559 x 330	117	9 1/2	241	14 x 11	356 x 279	238	5 1/2	140
22 x 12	559 x 305	126	9 1/2	241	14 x 10	356 x 254	261	5 1/2	140
22 x 11	559 x 279	138	9 1/2	241	14 x 9	356 x 229	291	5 1/2	140
22 x 10	559 x 254	152	9 1/2	241	14 x 8	356 x 203	327	5 1/2	140
					14 x 7	356 x 178	374	5 1/2	140
20 x 14	508 x 356	121	8 1/2	216	12 x 10	305 x 254	320	4 1/2	114
20 x 13	508 x 330	132	8 1/2	216	12 x 9	305 x 229	355	4 1/2	114
20 x 12	508 x 305	141	8 1/2	216	12 x 8	305 x 203	400	4 1/2	114
20 x 11	508 x 279	154	8 1/2	216	12 x 7	305 x 178	457	4 1/2	114
20 x 10	508 x 254	170	8 1/2	216	12 x 6	305 x 152	533	4 1/2	114
20 x 9	508 x 229	189	8 1/2	216					
18 x 14	457 x 356	137	7 1/2	191	11 x 8	279 x 203	450	4	102
18 x 13	457 x 330	148	7 1/2	191	11 x 7	279 x 178	515	4	102
18 x 12	457 x 305	160	7 1/2	191					
18 x 11	457 x 279	175	7 1/2	191	10 x 8	254 x 203	515	3 1/2	89
18 x 10	457 x 254	192	7 1/2	191	10 x 7	254 x 178	588	3 1/2	89
18 x 9	457 x 229	213	7 1/2	191	10 x 6	254 x 152	686	3 1/2	89

TABLE 9.5 ▪ Wood shingle coverage

SHINGLE COVERAGE

SHINGLES	LENGTH OF NO. 1 SHINGLES IN INCHES (mm)	APPROXIMATE COVERAGE IN SQ. FT. (m²) OF ONE SQUARE (4 BUNDLES) OF SHINGLES BASED ON FOLLOWING WEATHER EXPOSURES IN INCHES (mm)								
		3 1/2" (89)	4" (102)	4 1/2" (115)	5" (127)	5 1/2" (140)	6" (152)	6 1/2" (165)	7" (178)	7 1/2" (191)
NO.1	16" (406)	70 (6.50)	80 (7.43)	90 (8.36)	100* (9.29)	—	—	—	—	—
	18" (457)	—	72.5 (6.74)	81.5 (7.57)	90.5 (8.40)	100* (9.29)	—	—	—	—
	24" (610)	—	—	—	—	73.5 (6.83)	80 (7.43)	86.5 (8.04)	93 (8.64)	100* (9.29)

* MAXIMUM EXPOSURE RECOMMENDED FOR ROOFS

TABLE 9.6 ■ Wood shingle shapes and sizes

WOOD SHINGLES

NAME	LENGTH, THICKNESS, AND REFERENCE NOMENCLATURE	DESCRIPTION
NO.1** TAPER-SAWN	18" x .40" FIVEX (406mm x 10mm) 18" x .45" PERFECTIONS (457mm x 11mm) 24" x .50" ROYALS (610mm x 13mm)	TOP GRADE (NO. 1**) WOOD SHINGLES FOR USE AS ROOFING. THE WOOD BLANKS ARE RUN DIAGONALLY THROUGH A BANDSAW TO PRODUCE SHINGLES THAT ARE SAWN BOTH SIDES.
NO.1** FANCY SAWN BUTT	16" x .40" (406mm x 10mm) 18" x .45" (457mm x 11mm) 24" x .50" (610mm x 13mm)	TOP GRADE (NO. 1**) WOOD SHINGLE FOR USE AS ROOFING. TAPER-SAWN BOTH SIDES, WITH BUTT END CUT TO SPECIFIC SHAPE. A VARIETY OF FANCY BUTTS ARE AVAILABLE.

OCTAGON

DIAGONAL   HALF COVE   HEXAGONAL   ROUND

DIAMOND   ARROW   FISH-SCALE   SQUARE

** LOWER GRADES (e.g., NO. 2 AND NO. 3) ARE AVAILABLE AND ARE USED FOR STARTER OR UNDERCOURSING.

TABLES 9.7-A AND 9.7-B ■ Shake coverage and exposure table

SHAKE COVERAGE

NO. 1 GRADE SHAKE TYPE, LENGTH, AND THICKNESS IN INCHES (mm)		APPROXIMATE COVERAGE IN SQ. FT. (m²) FOR 5 BUNDLES WHEN SHAKES ARE APPLIED WITH AN AVERAGE 1/2" (13mm) SPACING AT THE FOLLOWING WEATHER EXPOSURES, IN INCHES (mm)				
		5" (127)	5 1/2" (140)	7 1/2" (191)	8 1/2" (216)	10" (254)
SHAKES	HEAVIES 24" X 3/4" (610 x 19)	—	—	75(b) (6.96)	85 (7.90)	100(c) (9.29)
	MEDIUMS 24" X 1/2" (610 x13)	—	—	75(b) (6.96)	85 (7.90)	100(c) (9.29)
NO.1 HANDSPLIT & RESAWN	HEAVIES 18" X 3/4" (457 x 19)	—	55(b) (5.11)	75(c) (6.96)	—	—
	MEDIUMS 18" X 1/2" (457 x 13)	—	55(b) (5.11)	75(c) (6.96)	—	—
	24" X 5/8" (610 x 16)	—	—	75(b) (6.96)	85 (7.90)	100(c) (9.29)
NO.1 TAPER-SAWN	18" X 5/8" (457 x 16)	—	55(b) (5.11)	75(c) (6.96)	—	—

Shake Type	Dimensions				
NO.1 TAPER-SPLIT	24" X 1/2" (610 x 13)	—	—	75(b) (6.96)	85 (7.90)	100(c) (9.29)
NO.1 STRAIGHT-SPLIT	18" X 3/8" (457 x 10)	—	65(b) (6.04)	90(c) (8.36)	—	—
HANDSPLIT STARTER	24" X 3/8" (610 x 10)	50 (4.65)	—	75(b) (6.96)	—	—
15" STARTER COURSE	15" X 3/8" (381 x 10)	USE WITH SHAKES APPLIED NOT OVER 7 1/2" (191mm) WEATHER EXPOSURE				
NO.2	24" X 3/8" (610 x 10)	USE WITH SHAKES APPLIED NOT OVER 10" (254mm) WEATHER EXPOSURE				

(a) 5 BUNDLES MAY COVER 100 SQ. FT. (9.29m) WHEN USED AS STARTER COURSE AT 10" (254mm) WEATHER EXPOSURE; 7 BUNDLES MAY COVER 100 SQ. FT. (9.29m) WHEN USED AS STARTER COURSE AT 7 1/2" (191mm) WEATHER EXPOSURE.

(b) MAXIMUM RECOMMENDED WEATHER EXPOSURE FOR TRIPLE COVERAGE ROOF CONSTRUCTION.

(c) MAXIMUM RECOMMENDED WEATHER EXPOSURE FOR DOUBLE COVERAGE ROOF CONSTRUCTION.

(d) MAXIMUM RECOMMENDED WEATHER EXPOSURE.

NOTE - ALL DIMENSIONS ARE APPROXIMATE

TABLE 9.8 ■ Roofing materials

Roofing type	Minimum slope	Life years	Relative cost	Weight (pounds/100 sq ft)
Asphalt shingle	4	15–20	Low	200–300
Slate	5	100	High	750–4000
Wood shake	3	50	High	300
Wood shingle	3	25	Medium	150

TABLE 9.9 ■ Potential life spans for various types of roofing materials

Material	Expected life span
Asphalt shingles	15 to 30 years
Fiberglass shingles	20 to 30 years
Wood shingles	20 years
Wood shakes	50 years
Slate	Indefinite
Clay tiles	Indefinite
Copper	In excess of 35 years
Aluminum	35 years
Built-up roofing	5 to 20 years

Note: All estimated life spans depend on installation procedure, maintenance, and climatic conditions.

TABLE 9.10 ■ Approximate weights of roofing materials, based on 100 square feet of material installed

Type of roofing	Weight in pounds
Clay shingle tile	1000–2000
Clay Spanish tile	800–1500
Slate	600–1600
Asphalt shingles	130–325
Wood shingles	200–300

TABLE 9.11 ■ Roofing materials lowest permissible slope

Material	Slope
Asphalt or fiberglass shingle	4 in 12 slope
Roll roofing with exposed nails	3 in 12 slope
Roll roofing with concealed nails 3" head lap	2 in 12 slope
Double coverage half lap	1 in 12 slope
Lower slope: Treat as a flat roof. Use a continuous membrane system: either built up felt/asphalt with crushed stone or metal system with sealed or soldered seams.	

Roofing calculations are fairly simple once you understand the concepts. Figuring the area of walls is even easier, and that is our next task. Let's move into the next chapter and find the shortcuts for figuring walls.

chapter **10**

WALL SHEATHING AND SIDING

Wall sheathing and siding is easier to calculate than it is to install. There is not much to figuring wall area. The math is usually as simple as multiplying the width of the wall by the height of the wall. This is very simple math. For example, a wall that is 40 feet in length and 8 feet in height will have a coverage area of 320 square feet. This is about as complicated as most of the math for siding and sheathing gets.

Siding is typically sold in units of squares. One square is equal to 100 square feet. With this in mind, you are seeking to find the number of square feet of coverage that is needed when doing a take-off for siding.

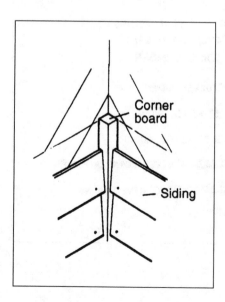

FIGURE 10.1 ■ Inside corner board trim

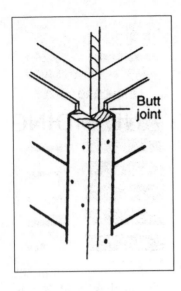

FIGURE 10.2 ■ Outside corner board trim

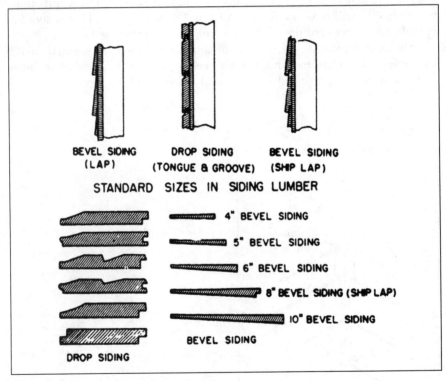

FIGURE 10.3 ■ Types of wood siding

An allowance has to be made for trim and waste when figuring the amount of siding needed for a job. Contractors often have their own ways of dealing with waste. And, the amount of waste can depend on the type of siding being used. Many contractors use a waste factor of 10-percent. Some contractors run their waste count as high as 15-percent. Since unused siding can usually be returned to a supplier, most contractors will order more siding than they expect to need. It is better to have additional siding on a job than it is to need to run out for more siding during a job.

SHEATHING

The type of sheathing used on buildings varies from region to region and is largely a matter of personal preference. Some regions use solid wood sheets of sheathing on the entire exterior wall areas. Other regions use rigid insulation boards for most of the exterior walls with wood sheets on the corners. All of the sheets normally used come in a 4' x 8' size and contain 32 square feet. This allows you to compute the total square footage of the wall space to be covered and divide it by 32 to determine the number of sheets of sheathing required to cover the wall.

Boards are sometimes used as wall sheathing. This is not common in modern construction, but it is still done from time to time. For example, there can be times that OSB board and plywood is hard to come by. During these times, some contractors use boards to maintain production schedules.

TABLE 10.1 ∎ Types and uses of plywood

Type	Use
Softwood veneer	Cross laminated plies or veneers—Sheathing, general construction and industrial use, etc.
Hardwood veneer	Cross laminated plies with hardwood face and back veneer—Furniture and cabinet work, etc.
Lumbercore plywood	Two face veneers and 2 crossband plies with an inner core of lumber strips—Desk and table tops, etc.
Medium-density overlay (MDO)	Exterior plywood with resin and fiber veneer—Signs, soffits, etc.
High-density overlay (HDO)	Tougher than MDO—Concrete forms, workbench tops, etc.
Plywood siding	T-111 and other textures used as one step sheathing and siding where codes allow.

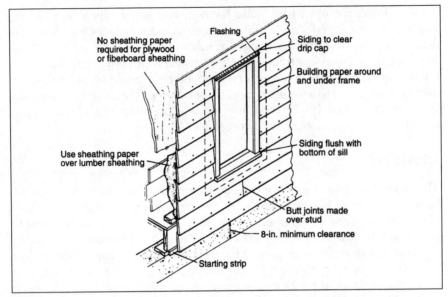

FIGURE 10.4 ■ **Detail of siding being installed over board sheathing**

TYPES OF SIDING

There are many types of siding to choose from. Some of the options available include:

- Vinyl siding
- Hardboard siding

TABLE 10.2 ■ **Siding comparison**

Material	Care	Life, yr	Cost
Aluminum	None	30	Medium
Hardboard	Paint Stain	30	Low
Horizontal wood	Paint Stain None	50+	Medium to high
Plywood	Paint Stain	20	Low
Shingles	Stain None	50+	High
Stucco	None	50+	Low to medium
Vertical wood	Paint Stain None	50+	Medium
Vinyl	None	30	Low

TABLE 10.3 ▪ Siding advantages and disadvantages

Material	Advantages	Disadvantages
Aluminum	Ease of installation over existing sidings Fire resistant	Susceptibility to denting, ratting in wind
Hardboard	Low cost Fast installation	Susceptibility to moisture in some
Horizontal wood	Good looks if of high quality	Slow installation Moisture/paint problems
Plywood	Low cost Fast installation	Short life Susceptibility to moisture in some
Shingles	Good looks Long life Low maintenance	Slow installation
Stucco	Long life Good looks in SW Low maintenance	Susceptibility to moisture
Vertical wood	Fast installation	Barn look if not of highest quality Moisture/paint problems
Vinyl	Low cost Ease of installation over existing siding	Fading of bright colors No fire resistance

- Cedar siding
- Pine siding
- Wood shake siding

There are pros and cons with different types of siding. Vinyl siding is popular due to its low cost, easy maintenance, and durability. Horizontal wood siding is also very popular. However, it is more expensive and takes longer to install than vinyl siding does. Hardboard siding is a good choice when a painted finish is wanted. This type of siding costs less than wood, installs quickly, and looks good once it is painted. One drawback is that hardboard siding cannot be stained.

▶ *trade* **tip**

When installing pine siding, be certain to stain or paint it before the siding becomes wet. If left untreated for long in wet conditions, pine siding is likely to turn black.

TABLE 10.4 ■ Siding compared

Material	Care	Life, yr	Cost
Aluminum	None	30	Medium
Hardboard	Paint Stain	30	Low
Horizontal wood	Paint Stain None	50+	Medium to high
Plywood	Paint Stain	20	Low
Shingles	Stain None	50+	High
Stucco	None	50+	Low to medium
Vertical wood	Paint Stain None	50+	Medium
Vinyl	None	30	Low

CALCULATING SIDING NEEDS

Calculating siding needs is simple enough. You are looking for the total square footage of area to cover. Straight walls are as easy as it gets. You just multiply the length by the height to obtain the total square footage of coverage.

Gable ends give some workers trouble when they are computing square footage. There is no need for this problem. Think of a gable as half of a straight wall. To determine the coverage area of a gable, measure the width of the base line. For example, on a house that is 24 feet wide, the base line of the gable will be 24 feet. Now go to the center of the base line and measure up to the peak of the gable. Assume that the distance is 16 feet. You now have the only two measurements needed to do the math. Multiply the base line width (24') by the height to the peak (16'). This will give you a total of 384 square feet. Since the gable is triangular, you have to divide the total by two in order to arrive at the coverage amount needed. The answer is 192 square feet. Math for sheathing and siding just isn't very complicated.

☑ *fast***facts**

Siding nails should be long enough to penetrate at least 1.5 inches into wall studs.

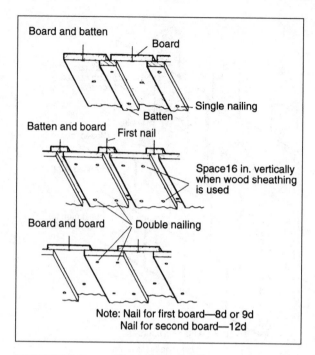

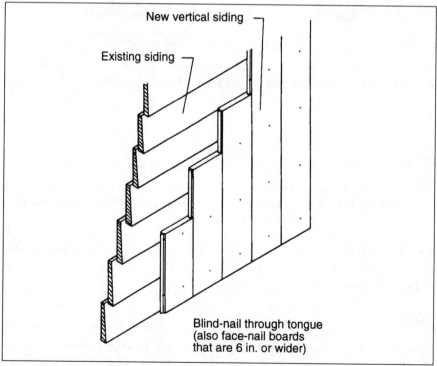

FIGURE 10.5 ■ **Applying board-and-batten siding**

FIGURE 10.6 ■ Installing vertical siding over existing horizontal siding

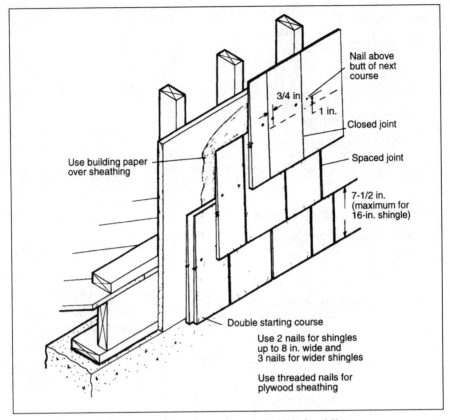

FIGURE 10.7 ▪ Single-course application of shingle siding

NAILING

The process of nailing siding is determined by the type of siding used. Choosing the proper nails and installing them in the correct manner is an important part of successful siding installations.

How many nails will you need with your siding order? How many nails are in a pound? You can check the description on a box of nails to see what you are getting. Reference tables are also handy for figuring the amount of nails needed for a job.

COVERAGE

The amount of coverage that different types of siding provide varies. Wooden sheets of siding give about 32 square feet of coverage per sheet. When it comes to other types of siding, the coverage per piece depends on the dimensions of the siding. Most suppliers will calculate the amount of

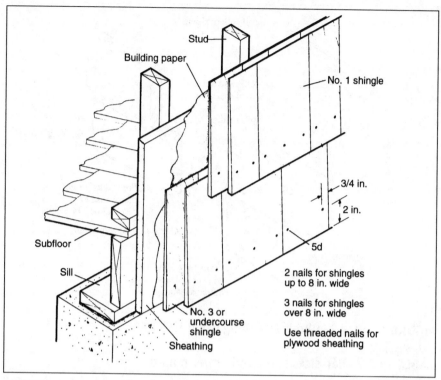

FIGURE 10.8 ▪ Double-course application of shingle siding

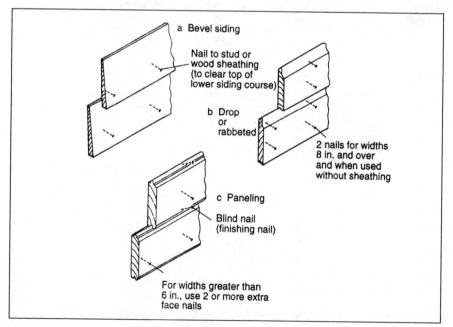

FIGURE 10.9 ▪ Details of nailing wood siding

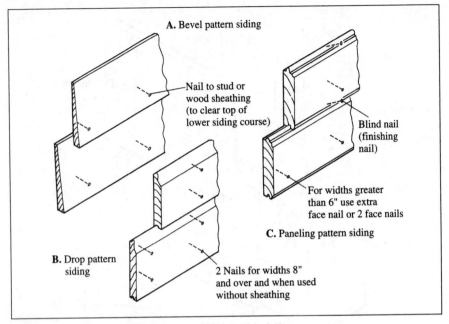

FIGURE 10.10 ■ **Methods of nailing wood siding**

TABLE 10.5 ■ **Nail sizes and number per pound**

Penny size "d"	Length	Approximate number per pound common	Approximate number per pound box	Approximate number per pound finish
2	1"	875	1000	1300
3	1¼"	575	650	850
4	1½"	315	450	600
5	1¾"	265	400	500
6	2"	190	225	300
7	2¼"	160		
8	2½"	105	140	200
9	2¾"	90		
10	3"	70	90	120
12	3¼"	60	85	110
16	3½"	50	70	90
20	4"	30	50	60
30	4½"	25		
40	5"	20		
50	5½"	15		
60	6"	10		

Note: Aluminum and c. c. nails are slightly smaller than other nails of the same penny size.

TABLE 10.6 ■ Screw lengths and available gauge numbers

Length	Guage Numbers
¼"	0 to 3
⅜"	2 to 7
½"	2 to 8
⅝"	3 to 10
¾"	4 to 11
⅞"	6 to 12
1"	6 to 14
1¼"	7 to 16
1½"	6 yo 18
1¾"	8 to 20
2"	8 to 20
2¼"	9 to 20
2½"	12 to 20
2¾"	14 to 20
3"	16 to 20
3½"	18 to 20
4"	18 to 20

TABLE 10.7 ■ Estimating the amount of nails needed per 100 square feet of bevel siding

Size	Pounds of Nails Needed
½ x 4	1½
½ x 5	1½
½ x 6	1
½ x 8	¾
⅝ x 8	¾
¾ x 8	¾
⅝ x 10	½
¾ x 10	½
¾ x 12	½

TABLE 10.8 ■ Estimating the amount of nails needed per 100 square feet of drop siding

Size	Pounds of Nails Needed
1 x 6	2.5
1 x 8	2

TABLE 10.9 ▪ Exposure distances for wood shingles and shakes on side walls

| | | Maximum exposure (in.) | | |
| | | | Double coursing | |
Material	Length of material (in.)	Single coursing	No. 1 grade	No. 2 grade
Shingles	16	7-1/2	12	10
	18	8-1/2	14	11
	24	11-1/2	16	14
Shakes	18	8-1/2	14	–
(handsplit	24	11-1/2	20	–
and resawn)	32	15	–	–

TABLE 10.10 ▪ Coverage estimator for bevel siding

Nominal size	Dress Width	Face Width	Area Factor
1 x 4	$5\frac{1}{2}$	$3\frac{1}{2}$	1.60
1 x 6	$5\frac{1}{2}$	$5\frac{1}{2}$	1.33
1 x 8	$7\frac{1}{4}$	$7\frac{1}{4}$	1.28
1 x 10	$9\frac{1}{4}$	$9\frac{1}{4}$	1.21
1 x 12	$11\frac{1}{4}$	$11\frac{1}{4}$	1.17

Note: Adjust coverage amounts to include the factors of trim and waste.

TABLE 10.11 ▪ Coverage estimator for shiplap siding

Nominal size	Dress Width	Face Width	Area Factor
1 x 6	$5\frac{7}{16}$	$4\frac{15}{16}$	1.22
1 x 8	$7\frac{1}{8}$	$8\frac{5}{8}$	1.21
1 x 10	$9\frac{1}{8}$	$6\frac{5}{8}$	1.16
1 x 12	$11\frac{1}{8}$	$10\frac{5}{8}$	1.13

Note: Adjust coverage amounts to include the factors of trim and waste.

TABLE 10.12 ▪ Coverage estimator for tongue-and-groove siding

Nominal size	Dress Width	Face Width	Area Factor
1 x 4	$3\frac{7}{16}$	$3\frac{3}{16}$	1.26
1 x 6	$5\frac{7}{16}$	$5\frac{3}{16}$	1.16
1 x 8	$7\frac{1}{8}$	$6\frac{7}{8}$	1.16
1 x 10	$9\frac{1}{8}$	$8\frac{7}{8}$	1.13
1 x12	$11\frac{1}{8}$	$10\frac{7}{8}$	1.10

Note: Adjust coverage amounts to include the factors of trim and waste.

siding needed for a job without charging their customers for the service. If you obtain the amount of square footage that needs to be sided, your supplier will probably do the math for you.

Now that we have completed the basic needs for figuring sheathing and siding, let's move to the next chapter and learn about windows.

chapter 11

WINDOWS AND EXTERIOR DOORS

Windows and exterior doors are primary parts of construction. These components are often expensive and frequently account for much of a building's appearance. As important as these elements are, there is minimal math involved with this phase of construction.

The key to success with windows and doors is in having accurate rough-in measurements. These figures should be obtained from the manufacturer of a product. To avoid costly, and embarrassing, problems during construction, take the time to get rough-in specifications for all products being installed before you finish the framing for these products. Even when using stock doors, it is best to have detailed rough-in specifications on hand.

DOORS

Doors come in a variety of shapes, sizes, and styles. The two most common exterior doors are panel doors and 9-lite doors. There are also terrace doors, sliding doors, and a host of other options.

TABLE 11.1 ■ Stock sizes of exterior doors

Height	Width
80"	36"
84"	36"
80"	34"
84"	34"
80"	32"
84"	32"

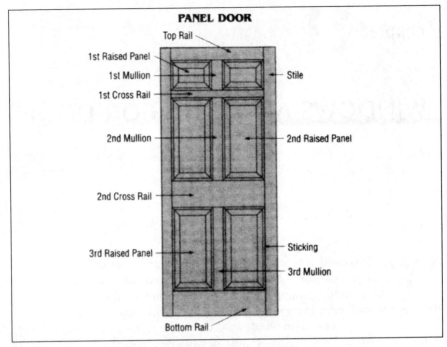

FIGURE 11.1 ■ Basic components of a panel door

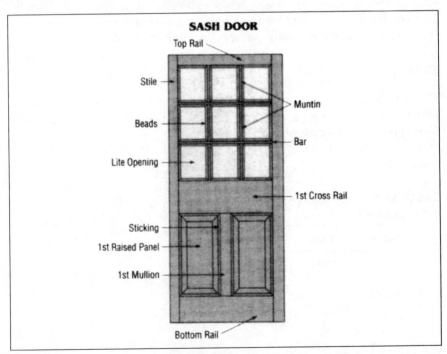

FIGURE 11.2 ■ Basic components of a door with glass lites

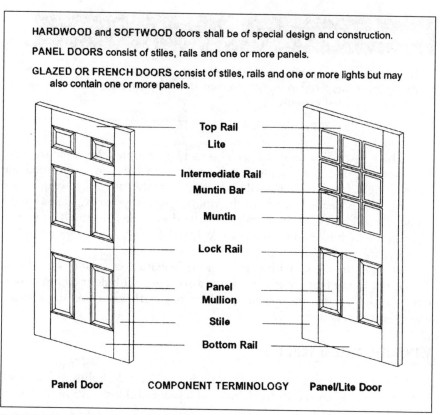

HARDWOOD and SOFTWOOD doors shall be of special design and construction.

PANEL DOORS consist of stiles, rails and one or more panels.

GLAZED OR FRENCH DOORS consist of stiles, rails and one or more lights but may also contain one or more panels.

Top Rail

Lite

Intermediate Rail

Muntin Bar

Muntin

Lock Rail

Panel

Mullion

Stile

Bottom Rail

Panel Door   COMPONENT TERMINOLOGY   Panel/Lite Door

FIGURE 11.3 ■ Door terminology

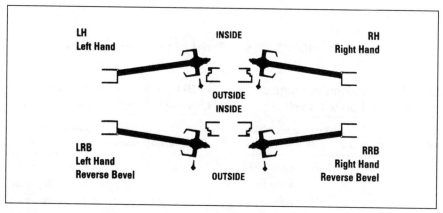

LH
Left Hand   INSIDE   RH
Right Hand

OUTSIDE

INSIDE

LRB
Left Hand   RRB
Reverse Bevel   Right Hand
                     Reverse Bevel
OUTSIDE

FIGURE 11.4 ■ Determining the swing of a door

▶ *trade* **tip**

Buy pre-hung doors that are pre-drilled for hardware. These doors cost more than basic components do, but the time saved on a job is well worth the added expense.

Which way will a door swing? This can be a problem for some workers. Not everyone understands how to tell where a left-hand door is needed and where a right-hand door is needed. I have a simple way of knowing which type of door is needed. Stand in the doorframe with your back against the frame where hinges will be installed. If the door swings in the direction of your right arm, it is a right-hand door. When the door is set to swing towards your left arm, it is a left-hand door.

All in all, there is very little need for mathematical calculations for doors. This is not to say that the math done in association with doors is not important. The point is that there is no need for fancy math when working with average door installations.

WINDOWS AND SKYLIGHTS

Like doors, the math requirements for windows and skylights are minimal. Rough openings are a critical element of a successful installation for a win-

• Double-hung
• Casement
• Fixed
• Awnings
• Sliding
• Skylight
• Bay
• Bow

FIGURE 11.5 ■ Types of residential windows

TABLE 11.2 ■ Maximum range of glass size, based on wind velocity

Wind velocity (mph)	Window glass thickness (in.) @ 0.133
30	64.5 square feet
40	32.25 square feet
65	16.1 square feet
120	4.5 square feet

TABLE 11.3 ■ Maximum range of glass size, based on wind velocity

Wind velocity (mph)	Window glass thickness (in.) @ 0.085
30	30 square feet
40	17.5 square feet
65	8.5 square feet
120	2.5 square feet

TABLE 11.4 ■ Heat gain for various types of glazing

Heat Gain and Performance Data

Heat Gain Data

In areas of the U.S. where cooling is the major energy cost, glazing may be the most important factor in energy-saving. That's because cooling costs are based almost solely on heat gains transmitted through the glass. The accompanying table is used to show maximum heat gain by type of glass.

Clear	Heat Gain	Tinted Grey/Bronze	Heat Gain	Medium Performance Reflective	Heat Gain
Single-pane ¼" or ⅛"	214	Single-pane grey ³⁄₁₆" (for comparison only)	165	Single-pane bronze (for comparison only)	106
Single-pane ³⁄₁₆" (for comparison only)	208	Single pane bronze ³⁄₁₆" (for comparison only)	168		
Double-pane (for comparison only)	186				
Double-pane high-performance insulating	113	Double-pane high-1 performance sun insulating			

☑ *fast***facts**

Some skylights are complete units that are designed to set on a roof, while others require framing a curb for the unit to sit above the roofline. Be sure to take this into consideration during the framing process.

dow or skylight. As discussed with doors, request rough-in specifications from manufacturers to determine your goal in rough openings. These requirements will vary from window type to window type, but manufacturers, and probably your supplier, should be happy to give you the measurements needed.

Since there is not much math to talk about here, let's move into the next chapter and talk about interior doors and trim.

chapter **12**

INTERIOR DOORS AND TRIM

There is not a lot of math involved when working with interior doors and trim. The type of math needed is not much more than reading a tape measure. Knowing how to cut wood on various angles is required, but a miter box and a combination square helps you to get this job done. Most carpenters use electric miter saws, often called chop saws, to minimize math and the time required for cutting trim.

Trim is calculated in linear feet, so there is little to discuss in regards to math for this portion of a job. If you are using trim kits for windows, they will be precut. There are many types of joints and angles involved with the installation of trim, but not much math.

TYPES OF TRIM

There are many types of interior trim to choose from. Crown molding is used to trim areas between a wall and ceiling. It is also used as trim for mantles over fireplaces. Cove molding covers the area between walls and ceilings and can be used as trim for inside corners. Quarter round is used to hide gaps between flooring and walls or baseboards. Corner bead is meant to cover outside corners. Half round hides joints between two butting pieces of paneling. Base molding covers the gap between floors and walls.

> ▶ *trade* **tip**
>
> Joints that are made in trim can be strengthened with the use of a wood biscuit that is glued into place. This procedure requires the use of a biscuit joiner machine.

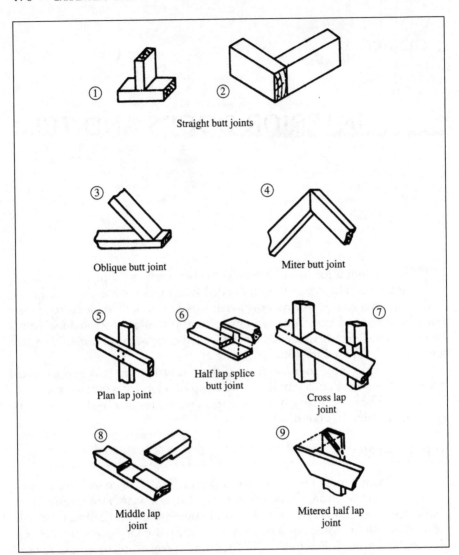

FIGURE 12.1 ▪ **Butt and lap joints**

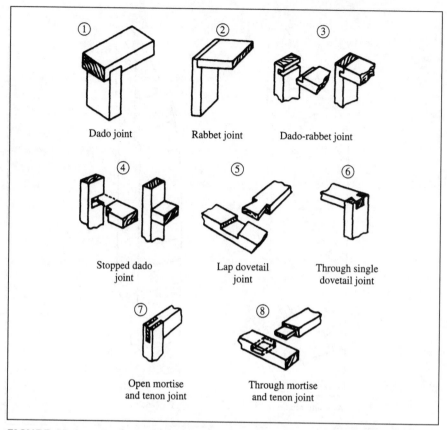

FIGURE 12.2 ■ Dado, rabbet, dovetail, and mortise and tenon joints

- Crown
- Rabbeted half round
- Half round
- Corner bead
- Sliding door
- Handrail
- Cove
- Quarter round
- Dowel
- Picture rail
- Scoop
- Edge

FIGURE 12.3 ■ Types of trim molding

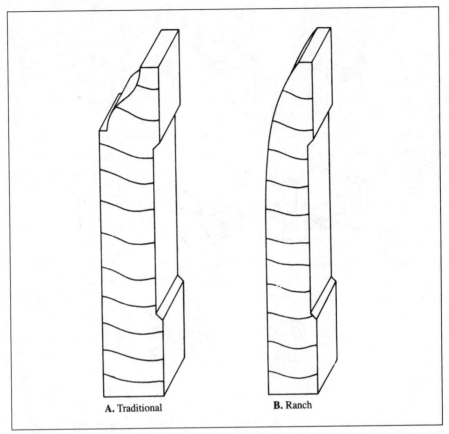

A. Traditional   B. Ranch

FIGURE 12.4 ■ Base molding

INTERIOR DOORS

Interior doors, like trim, don't require a lot of math. Most carpenters buy pre-hung doors. This means that the jamb and trim is shipped as a part of the door package. How easy is that? There are, however, rough opening sizes that you must be aware of. These measurements should come from the manufacturer of the products that you will be installing.

▶ *trade* **tip**

Never use fingerjoint trim if the trim is to be stained. Fingerjoint trim is fine when it is painted, but it will have a poor appearance if it is stained.

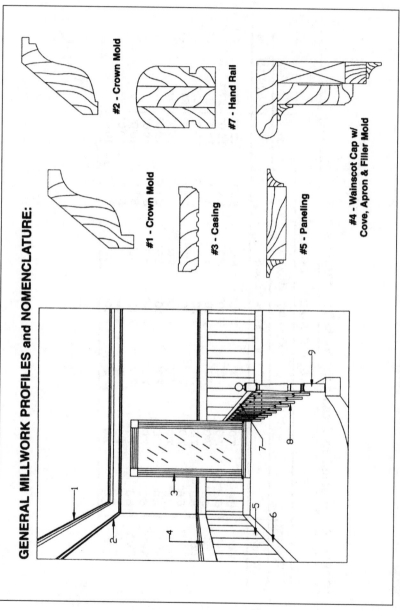

FIGURE 12.5 ■ General millwork profiles and uses

TABLE 12.1 ■ Standard opening size for hollow metal doors

	STANDARD OPENING SIZE								
Opening	Opening Heights								
Widths	1 3/4 " Doors							1 3/8 " Doors	
2'0"	6'8"	7'0"	7'2"	7'10"	8'0"	8'10"	10'0"	6'8"	7'0"
2'4"	6'8"	7'0"	7'2"	7'10"	8'0"	8'10"	10'0"	6'8"	7'0"
2'6"	6'8"	7'0"	7'2"	7'10"	8'0"	8'10"	10'0"	6'8"	7'0"
2'8"	6'8"	7'0"	7'2"	7'10"	8'0"	8'10"	10'0"	6'8"	7'0"
2'10"	6'8"	7'0"	7'2"	7'10"	8'0"	8'10"	10'0"	6'8"	7'0"
3'0"	6'8"	7'0"	7'2"	7'10"	8'0"	8'10"	10'0"	6'8"	7'0"
3'4"	6'8"	7'0"	7'2"	7'10"	8'0"	8'10"	10'0"		
3'6"	6'8"	7'0"	7'2"	7'10"	8'0"	8'10"	10'0"		
3'8"	6'8"	7'0"	7'2"	7'10"	8'0"	8'10"	10'0"		
3'10"	6'8"	7'0"	7'2"	7'10"	8'0"	8'10"	10'0"		
4'0"	6'8"	7'0"	7'2"	7'10"	8'0"	8'10"	10'0"		

TABLE 12.2 ■ Measurements for sliding-glass doors

Glass size (in inches)	Frame size (width × height, feet and inches)	Rough opening (width × height, feet and inches)
33 × 76¾	6-0 × 6-10¾	6-0½ × 6-11¼
45 × 76¾	8-0 × 6-10¾	8-0½ × 6-11¼
57 × 76¾	10-0 × 6-10¾	10-0½ × 6-11¼
33 × 76¾	9-0 × 6-10¾	8-0½ × 6-11¼
45 × 76¾	12-0 × 6-10¾	12-0½ × 6-11¼
57 × 76¾	15-0 × 6-10¾	15-0½ × 6-11¼
33 × 76¾	11-11 × 6-10¾	11-11½ × 6-11¼
45 × 76¾	15-11 × 6-10¾	15-11½ × 6-11¼
57 × 76¾	19-11 × 6-10¾	19-11½ × 6-11¼

TABLE 12.3 ■ Stock sizes of interior doors

Height	Width
80"	24"
80"	28"
84"	28"
80"	30"
84"	30"
80"	32"
84"	32"

Rough openings vary with the types of doors being installed. Doors can be made to size, but most jobs are done by using stock door sizes. See Table 12.3 for some sample stock sizes of interior doors.

Another consideration when dealing with interior doors and passageways is the minimum width requirements. Check your local code to establish the required minimum standards in your region.

TABLE 12.4 ■ Widths of passageways

Passageway	Recommended	Minimum
Stairs	40"	36"
Landings	40"	36"
Main hall	48"	36"
Minor hall	36"	30"
Interior door	32"	28"
Exterior door	36"	36"

TABLE 12.5 ■ Sample specifications for two-panel bifold doors

Size	Door Width	Opening Width	Finished Heights
Sample Specifications for 6' 8" Two-Panel Bifold Doors.			
1' 6"	17½"	18½"	6' 8¾"
2' 0"	23½"	24½"	6' 8¾"
2' 6"	29½"	30½"	6' 8¾"
3' 0"	35½"	36½"	6' 8¾"
Sample Specifications for 7' 6" Two-Panel Bifold Doors.			
1' 6"	17½"	18½"	6' 7' 5¼"
2' 0"	23½"	24½"	6' 7' 5¼"
2' 6"	29½"	30½"	6' 7' 5¼"
3' 0"	35½"	36½"	6' 7' 5¼"
Sample Specifications for 8' Two-Panel Bifold Doors.			
1' 6"	17½"	18½"	6' 7' 11¼"
2' 0"	23½"	24½"	6' 7' 11¼"
2' 6"	29½"	30½"	6' 7' 11¼"
3' 0"	35½"	36½"	6' 7' 11¼"

TABLE 12.6 ■ Sample specifications for four-panel bifold doors

Size	Door Width	Opening Width	Finished Heights
Sample Specifications for 6' 8" Four-Panel Bifold Doors.			
3' 0"	35"	36"	6' 8¾"
4' 0"	47"	48"	6' 8¾"
5' 0"	59"	60"	6' 8¾"
6' 0"	71"	72"	6' 8¾"
Sample Specifications for 7' 6" Four-Panel Bifold Doors.			
3' 0"	35"	36"	7' 5¼"
4' 0"	47"	48"	7' 5¼"
5' 0"	59"	60"	7' 5¼"
6' 0"	71"	72"	7' 5¼"
Sample Specifications for 8' Four-Panel Bifold Doors.			
3' 0"	35"	36"	7' 11¼"
4' 0"	47"	48"	7' 11¼"
5' 0"	59"	60"	7' 11¼"
6' 0"	71"	72"	7' 11¼"

☑ *fast***facts**

Rough openings for doors are usually framed to be 3 inches taller than the dimensions of a door to be installed. The width of the opening is typically two and one-half inches wider than the actual door dimensions.

Always request manufacturer's specifications for the products that you will be installing. Confirm all rough openings well in advance of installing doors. You don't want to wait until the last minute to discover that the rough openings are sized improperly.

Since there is little math to discuss in this chapter, let's move to Chapter 13 and explore the needs for math when working with cabinets and counters.

CABINETS AND COUNTERS

Cabinets and counters don't require any mystical math. The mathematical needs for this phase of construction are typically simple, linear measurements. Established standards are typically used to establish the height of a countertop. The same is true for setting wall cabinets. There can be, however, deviations from standards. For example, you may encounter a customer who is very tall and who wants a higher countertop. The reverse could be true with a customer who is not as tall as some people. While setting standards is normal, the best option a contractor has is to determine desired heights from customers.

In addition to establishing dimensions that will satisfy your customers, you should determine the type of cabinets wanted by customers. The decision made by a customer can affect your costs considerably. Pin down specifics on cabinets before you bid a job.

COMMON CABINET FACTS

Since we don't have any substantial math to discuss for cabinets or counters, we will end this chapter with a graphic tour of common cabinet features and dimensions. Then you can move into Chapter 14 to reap the benefits of conversion tables of all sorts. For now, check out the following illustrations to get an idea of the types of measurements to have in mind when planning a cabinet job.

▶ *trade* **tip**

If your customer wants custom cabinets, allow plenty of time for the construction of cabinets. Many customers will be happy with stock cabinets, but there are always customers who want a custom job. This can add weeks to the time required to complete a job.

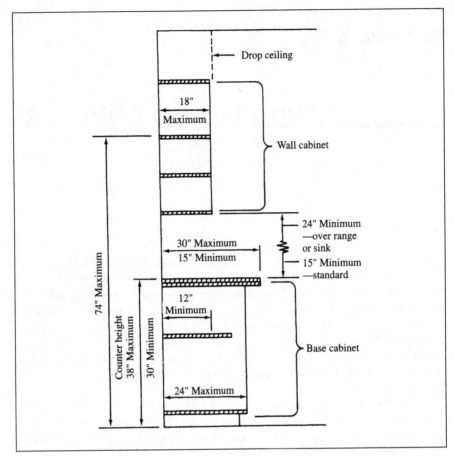

FIGURE 13.1 ▪ **Common kitchen cabinet dimensions**

TABLE 13.1 ▪ Features of kitchen cabinets

Type of cabinet	Features
Steel	Noisy
	Might rust
	Poor resale value
Hardwood	Sturdy
	Durable
	Easy to maintain
	Excellent resale value
Hardboard	Sturdy
	Durable
	Easy to maintain
	Good resale value
Particleboard	Sturdy
	Normally durable
	Easy to maintain
	Fair resale value

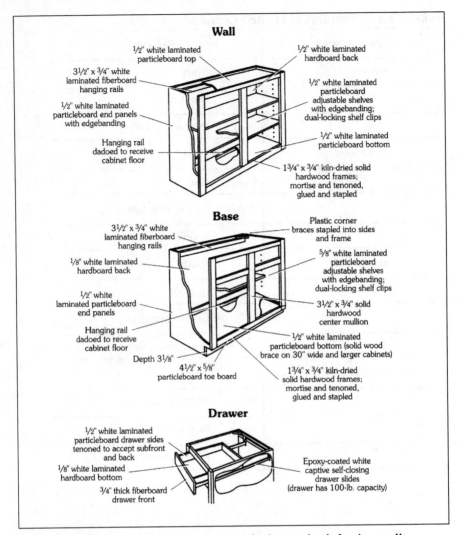

Wall

1/2" white laminated particleboard top

1/2" white laminated hardboard back

3½" x 3/4" white laminated fiberboard hanging rails

1/2" white laminated particleboard adjustable shelves with edgebanding; dual-locking shelf clips

1/2" white laminated particleboard end panels with edgebanding

1/2" white laminated particleboard bottom

Hanging rail dadoed to receive cabinet floor

13/4" x 3/4" kiln-dried solid hardwood frames; mortise and tenoned, glued and stapled

Base

3½" x 3/4" white laminated fiberboard hanging rails

Plastic corner braces stapled into sides and frame

1/8" white laminated hardboard back

5/8" white laminated particleboard adjustable shelves with edgebanding; dual-locking shelf clips

1/2" white laminated particleboard end panels

3½" x 3/4" solid hardwood center mullion

Hanging rail dadoed to receive cabinet floor

1/2" white laminated particleboard bottom (solid wood brace on 30" wide and larger cabinets)

Depth 31/8"

4½" x 5/8" particleboard toe board

13/4" x 3/4" kiln-dried solid hardwood frames; mortise and tenoned, glued and stapled

Drawer

1/2" white laminated particleboard drawer sides tenoned to accept subfront and back

1/8" white laminated hardboard bottom

Epoxy-coated white captive self-closing drawer slides (drawer has 100-lb. capacity)

3/4" thick fiberboard drawer front

FIGURE 13.2 ■ Detailed cross-section of what to look for in quality cabinets (Courtesy of Wellborn Cabinet, Inc.)

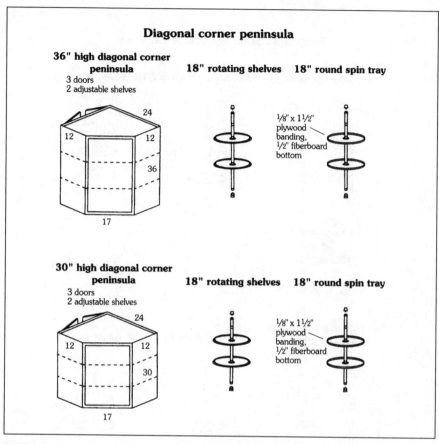

FIGURE 13.3 ■ Details for appliance cabinets, wall cabinets, microwave cabinets, and return-angle wall cabinets (Courtesy of Wellborn Cabinet, Inc.)

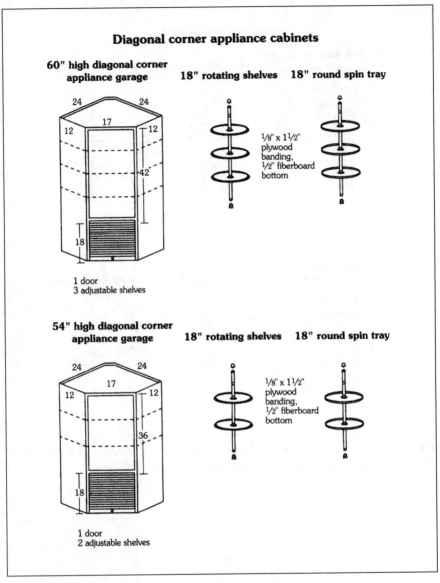

FIGURE 13.4 ■ Diagonal corner appliance cabinets

(Courtesy of Wellborn Cabinet, Inc.)

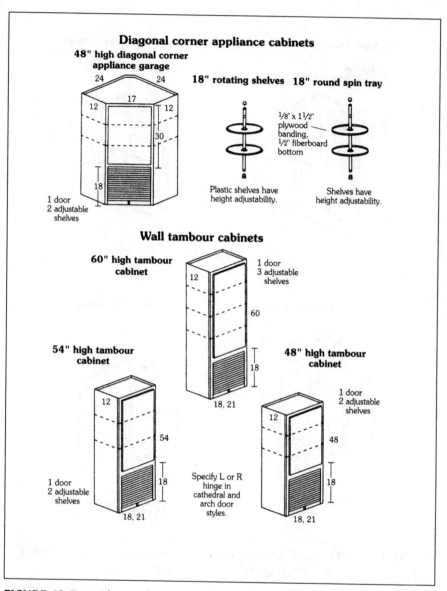

Diagonal corner appliance cabinets

48" high diagonal corner appliance garage

18" rotating shelves   18" round spin tray

Plastic shelves have height adjustability.

Shelves have height adjustability.

Wall tambour cabinets

60" high tambour cabinet

54" high tambour cabinet

48" high tambour cabinet

Specify L or R hinge in cathedral and arch door styles.

FIGURE 13.5 ▪ **Diagonal corner peninsula cabinets that give customers superior storage in a small space** (Courtesy of Wellborn Cabinet, Inc.)

42" high microwave cabinet

2 doors
1 adjustable shelf
1 17¼" deep removable shelf.

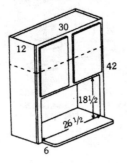

36" high microwave cabinet

2 doors
1 17¼" deep removable shelf.

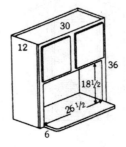

48" high microwave cabinet

2 doors
1 adjustable shelf
1 17¼" deep removable shelf.

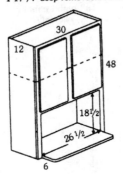

Base/wall return angle

2 doors
2 fixed shelves

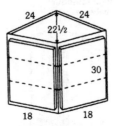

Inside angle 140°.
Outside angle 220°.
Square doors standard.
Ends not finished.

FIGURE 13.6 ■ Diagonal corner peninsula cabinets that give customers superior storage in a small space (Courtesy of Wellborn Cabinet, Inc.)

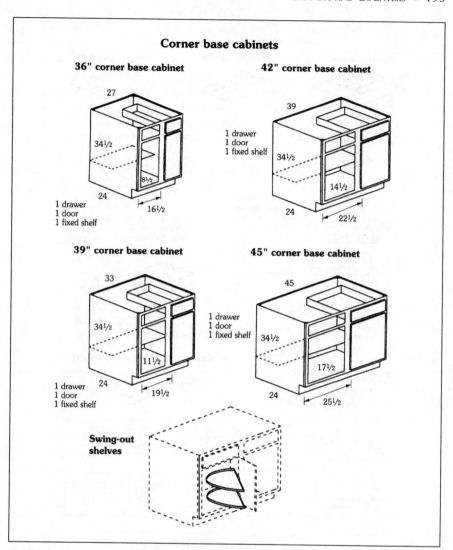

Corner base cabinets

36" corner base cabinet

27
34½
8½
24
16½
1 drawer
1 door
1 fixed shelf

42" corner base cabinet

39
1 drawer
1 door
1 fixed shelf
34½
14½
24
22½

39" corner base cabinet

33
34½
11½
24
19½
1 drawer
1 door
1 fixed shelf

45" corner base cabinet

45
1 drawer
1 door
1 fixed shelf
34½
17½
24
25½

Swing-out shelves

FIGURE 13.7 ■ **Options for corner base cabinets**
(Courtesy of Wellborn Cabinet, Inc.)

Base accessories

Silverware tray

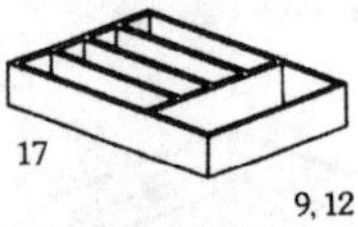

17

9, 12

Silverware divider

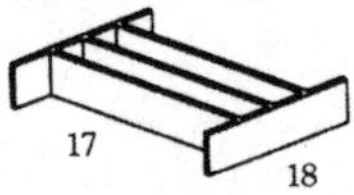

17

18

Drawer spice rack

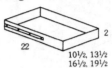

22

20

White high-density polyethylene.
May be trimmed.

Silverware tray

Double-tiered white tray. Guides have epoxy-coated
sides. Top drawer integrates with bottom drawer by
sliding back to reveal bottom drawer. Side flanges
may be trimmed. Back of drawer must be removed
for tray to be functional.

Interior drawer		
Min. height	Width	Min. depth
3½"	12¼ to 14¾"	17¼"

Utensil drawer kit

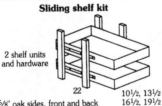

18½   13½

3½

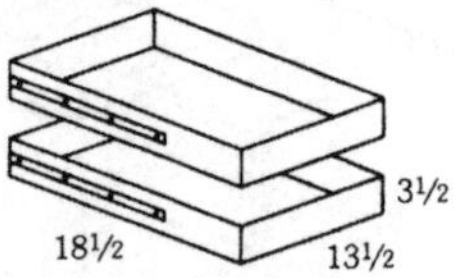

Slides are ³⁄₈" oak. Bottom is ³⁄₁₆"
plywood. Kit includes 2 shelves and
guides for installation.

Sliding shelf kit

2 shelf units
and hardware

22

10½, 13½
16½, 19½

⅝" oak sides, front and back
³⁄₁₆" hardwood bottom.
Adjustable wooden mounting brackets.

Single roll-out shelf

22

2

10½, 13½
16½, 19½

One shelf and guides for installation.
½" laminated particleboard sides.
¼" oak front. Bottom is ⅛" hardboard.

Cutting board

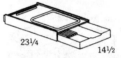

23¼

14½

Includes cutting board, knife divider and drawer.
Cutting board will replace existing drawer.

Double-tiered cutlery/cutting board

Includes cutting board and cutlery divider.
Furnished with two hinges and necessary
screws for installation. White cutting board is
made of non-skid, non-absorbent polystyrene
and is removable. Dishwasher safe. Side flanges
may be trimmed.

Interior drawer		
Min. height	Width	Min. depth
3½"	18¼ to 20¼"	17¼"
3½"	12¼ to 14¼"	17¼"

FIGURE 13.8 ▪ Options for corner base cabinets

(Courtesy of Wellborn Cabinet, Inc.)

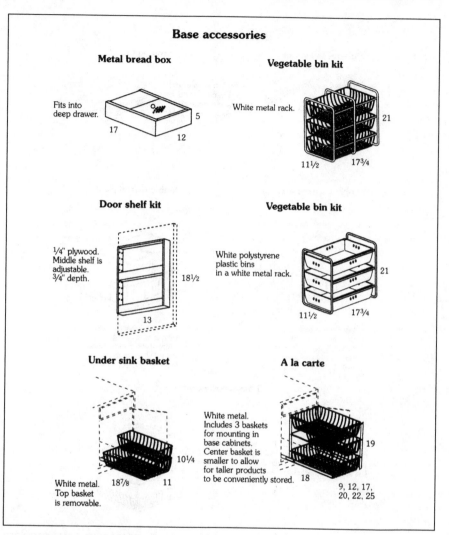

Base accessories

Metal bread box

Fits into
deep drawer.

17   5
    12

Vegetable bin kit

White metal rack.

21

11½   17¾

Door shelf kit

¼" plywood.
Middle shelf is
adjustable.
¾" depth.

18½

13

Vegetable bin kit

White polystyrene
plastic bins
in a white metal rack.

21

11½   17¾

Under sink basket

10¼

White metal.   18⅞   11
Top basket
is removable.

A la carte

White metal.
Includes 3 baskets
for mounting in
base cabinets.
Center basket is
smaller to allow
for taller products
to be conveniently stored.   18

19

9, 12, 17,
20, 22, 25

FIGURE 13.9 ■ Options for corner base cabinets
(Courtesy of Wellborn Cabinet, Inc.)

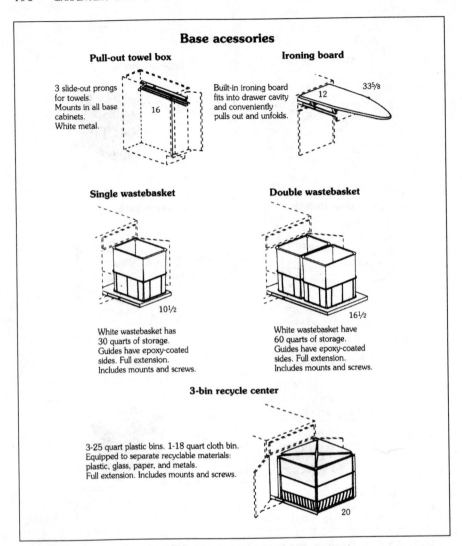

Base acessories

Pull-out towel box

3 slide-out prongs for towels. Mounts in all base cabinets. White metal.

16

Ironing board

Built-in ironing board fits into drawer cavity and conveniently pulls out and unfolds.

12

33⅝

Single wastebasket

10½

White wastebasket has 30 quarts of storage. Guides have epoxy-coated sides. Full extension. Includes mounts and screws.

Double wastebasket

16½

White wastebasket have 60 quarts of storage. Guides have epoxy-coated sides. Full extension. Includes mounts and screws.

3-bin recycle center

3-25 quart plastic bins. 1-18 quart cloth bin. Equipped to separate recyclable materials: plastic, glass, paper, and metals. Full extension. Includes mounts and screws.

20

FIGURE 13.10 ■ Options for corner base cabinets
(Courtesy of Wellborn Cabinet, Inc.)

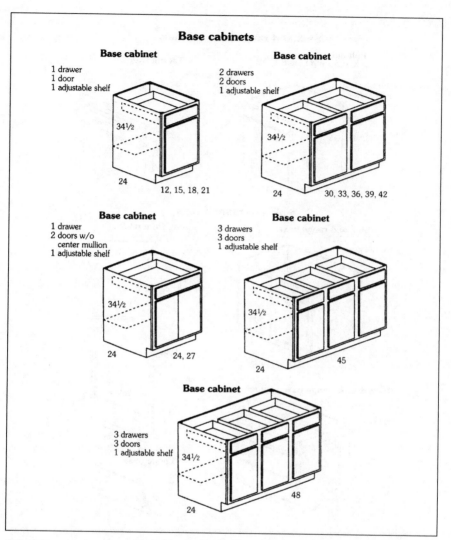

FIGURE 13.11 ■ Options for corner base cabinets
(Courtesy of Wellborn Cabinet, Inc.)

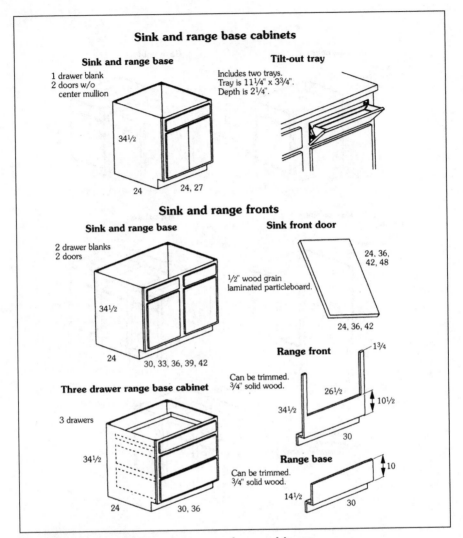

FIGURE 13.12 ■ Options for corner base cabinets
(Courtesy of Wellborn Cabinet, Inc.)

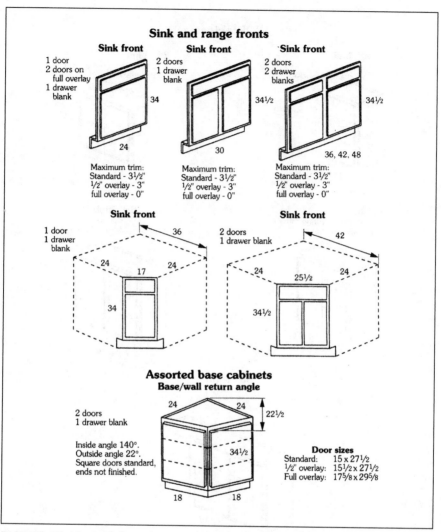

FIGURE 13.13 ▪ **Sink and range fronts** (Courtesy of Wellborn Cabinet, Inc.)

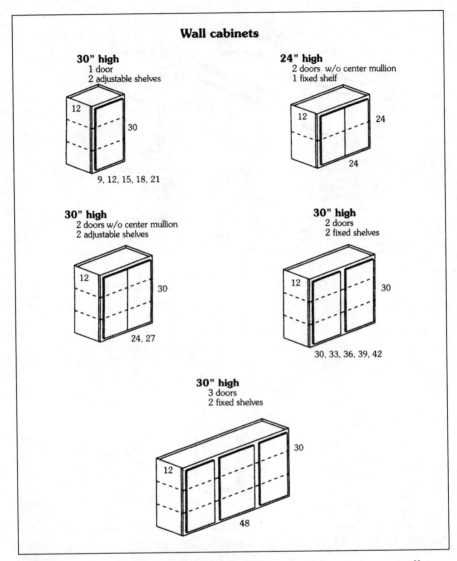

Wall cabinets

30" high
1 door
2 adjustable shelves

12 · 30

9, 12, 15, 18, 21

24" high
2 doors w/o center mullion
1 fixed shelf

12 · 24

24

30" high
2 doors w/o center mullion
2 adjustable shelves

12 · 30

24, 27

30" high
2 doors
2 fixed shelves

12 · 30

30, 33, 36, 39, 42

30" high
3 doors
2 fixed shelves

30

12

48

FIGURE 13.14 ■ **Cabinet specifications for wall cabinets, corner wall cabinets, and diagonal wall cabinets** (Courtesy of Wellborn Cabinet, Inc.)

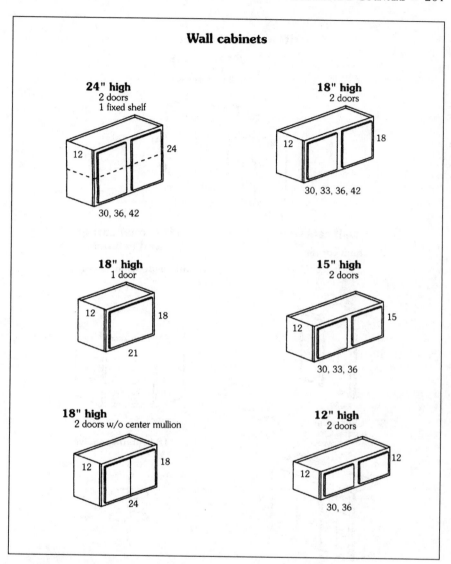

FIGURE 13.15 ▪ Cabinet specifications for wall cabinets
(Courtesy of Wellborn Cabinet, Inc.)

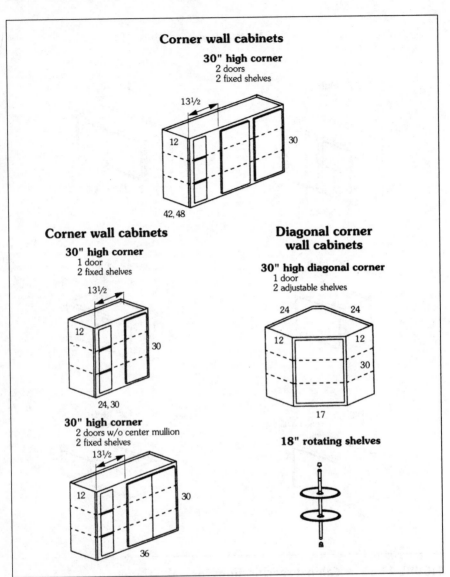

FIGURE 13.16 ■ Cabinet specifications for wall cabinets
(Courtesy of Wellborn Cabinet, Inc.)

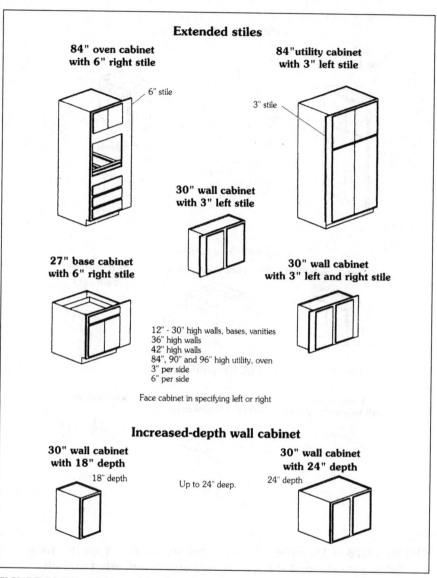

Extended stiles

84" oven cabinet with 6" right stile

6" stile

84"utility cabinet with 3" left stile

3" stile

30" wall cabinet with 3" left stile

27" base cabinet with 6" right stile

30" wall cabinet with 3" left and right stile

12" - 30" high walls, bases, vanities
36" high walls
42" high walls
84", 90" and 96" high utility, oven
3" per side
6" per side

Face cabinet in specifying left or right

Increased-depth wall cabinet

30" wall cabinet with 18" depth

18" depth

Up to 24" deep.

30" wall cabinet with 24" depth

24" depth

FIGURE 13.17 ■ **Examples of customized stock cabinet specifications that allow for extended stiles, increased depth, and other options**
(Courtesy of Wellborn Cabinet, Inc.)

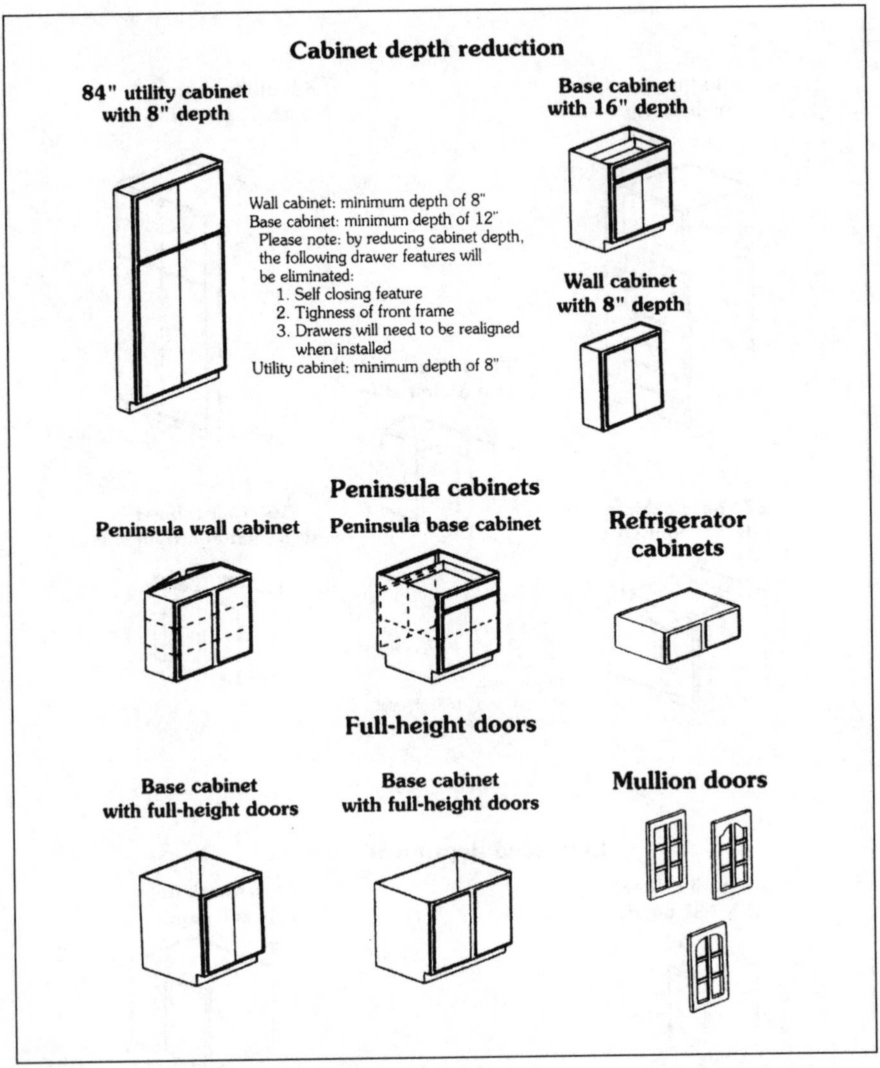

Cabinet depth reduction

84" utility cabinet with 8" depth

Base cabinet with 16" depth

Wall cabinet: minimum depth of 8"
Base cabinet: minimum depth of 12"
Please note: by reducing cabinet depth,
the following drawer features will
be eliminated:
    1. Self closing feature
    2. Tighness of front frame
    3. Drawers will need to be realigned
       when installed
Utility cabinet: minimum depth of 8"

Wall cabinet with 8" depth

Peninsula cabinets

Peninsula wall cabinet

Peninsula base cabinet

Refrigerator cabinets

Full-height doors

Base cabinet with full-height doors

Base cabinet with full-height doors

Mullion doors

FIGURE 13.18 ■ Examples of customized stock cabinet specifications that allow for extended stiles, increased depth, and other options
(Courtesy of Wellborn Cabinet, Inc.)

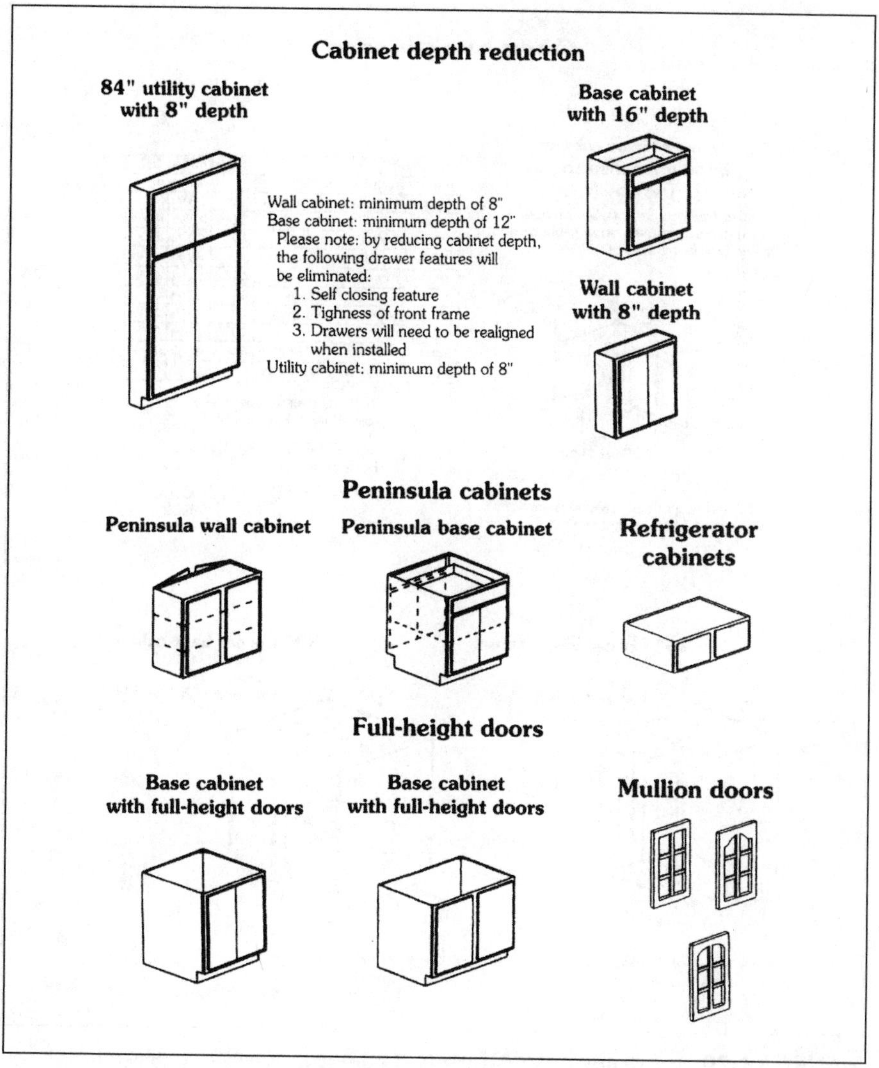

Cabinet depth reduction

84" utility cabinet with 8" depth

Base cabinet with 16" depth

Wall cabinet: minimum depth of 8"
Base cabinet: minimum depth of 12"
  Please note: by reducing cabinet depth,
  the following drawer features will
  be eliminated:
      1. Self closing feature
      2. Tighness of front frame
      3. Drawers will need to be realigned
          when installed
Utility cabinet: minimum depth of 8"

Wall cabinet with 8" depth

Peninsula cabinets

Peninsula wall cabinet   **Peninsula base cabinet**   **Refrigerator cabinets**

Full-height doors

Base cabinet with full-height doors   **Base cabinet with full-height doors**   **Mullion doors**

FIGURE 13.19 ■ **Corner sink base cabinets** (Courtesy of Wellborn Cabinet, Inc.)

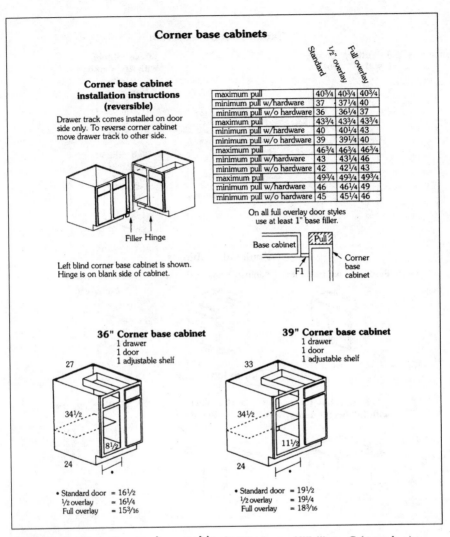

Corner base cabinets

Corner base cabinet installation instructions (reversible)

Drawer track comes installed on door side only. To reverse corner cabinet move drawer track to other side.

Filler Hinge

Left blind corner base cabinet is shown. Hinge is on blank side of cabinet.

	Standard	½" overlay	Full overlay
maximum pull	40¾	40¾	40¾
minimum pull w/hardware	37	37¼	40
minimum pull w/o hardware	36	36¼	37
maximum pull	43¾	43¾	43¾
minimum pull w/hardware	40	40¼	43
minimum pull w/o hardware	39	39¼	40
maximum pull	46¾	46¾	46¾
minimum pull w/hardware	43	43¼	46
minimum pull w/o hardware	42	42¼	43
maximum pull	49¾	49¾	49¾
minimum pull w/hardware	46	46¼	49
minimum pull w/o hardware	45	45¼	46

On all full overlay door styles use at least 1" base filler.

Base cabinet — Pull

Corner base cabinet

F1

36" Corner base cabinet
1 drawer
1 door
1 adjustable shelf

27

34½

8½

24

• Standard door = 16½
½ overlay = 16¼
Full overlay = 15³⁄₁₆

39" Corner base cabinet
1 drawer
1 door
1 adjustable shelf

33

34½

11½

24

• Standard door = 19½
½ overlay = 19¼
Full overlay = 18³⁄₁₆

FIGURE 13.20 ■ **Corner base cabinets** (Courtesy of Wellborn Cabinet, Inc.)

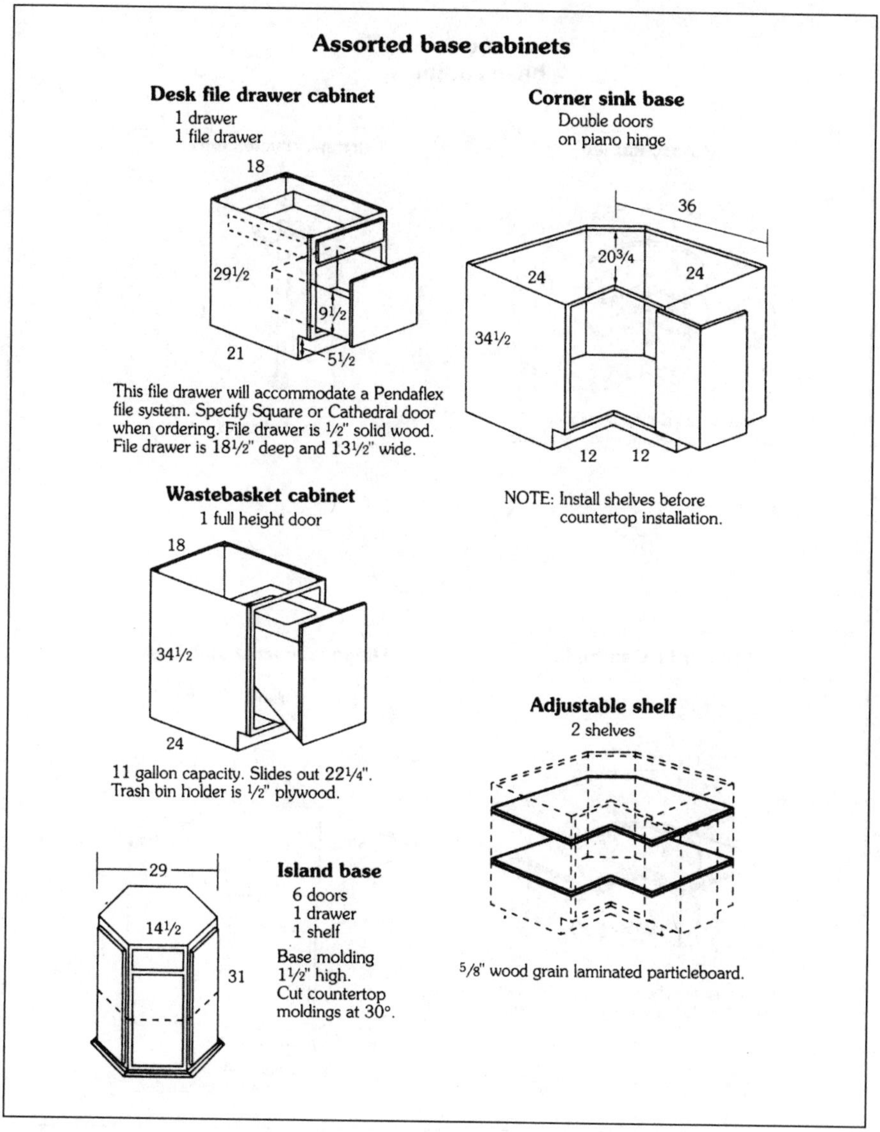

Assorted base cabinets

Desk file drawer cabinet
1 drawer
1 file drawer

18
29½
9½
21
5½

This file drawer will accommodate a Pendaflex file system. Specify Square or Cathedral door when ordering. File drawer is ½" solid wood. File drawer is 18½" deep and 13½" wide.

Corner sink base
Double doors
on piano hinge

36
20¾
24
24
34½
12 12

NOTE: Install shelves before countertop installation.

Wastebasket cabinet
1 full height door

18
34½
24

11 gallon capacity. Slides out 22¼".
Trash bin holder is ½" plywood.

Island base
6 doors
1 drawer
1 shelf

Base molding
1½" high.
Cut countertop
moldings at 30°.

29
14½
31

Adjustable shelf
2 shelves

⅝" wood grain laminated particleboard.

FIGURE 13.21 ■ Assorted base cabinets (Courtesy of Wellborn Cabinet, Inc.)

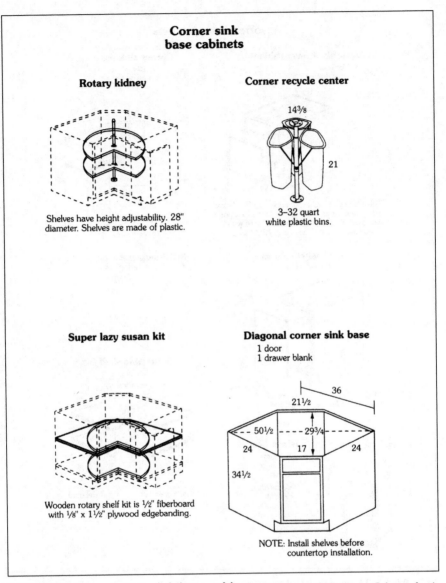

FIGURE 13.22 ■ **Corner sink base cabinets** (Courtesy of Wellborn Cabinet, Inc.)

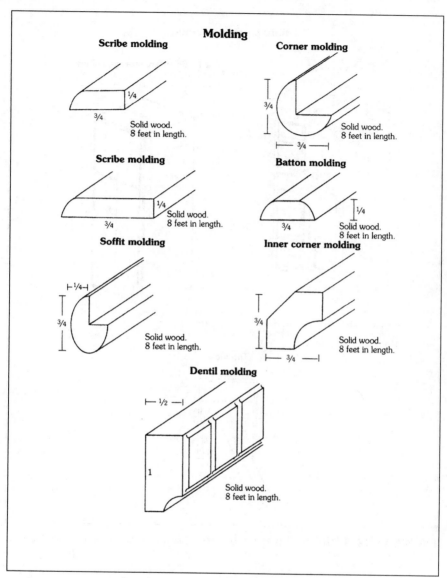

FIGURE 13.23 ▪ Types of moldings used with cabinets
(Courtesy of Wellborn Cabinet, Inc.)

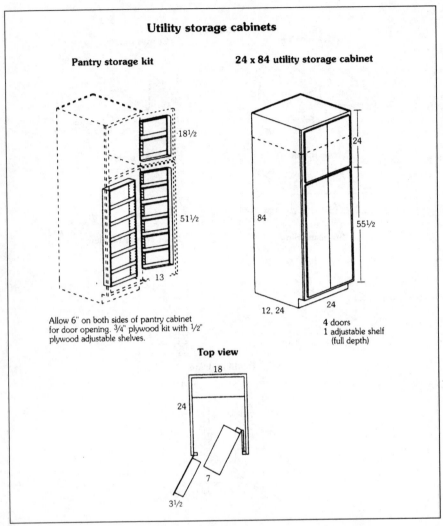

FIGURE 13.24 ■ **Utility storage cabinets** (Courtesy of Wellborn Cabinet, Inc.)

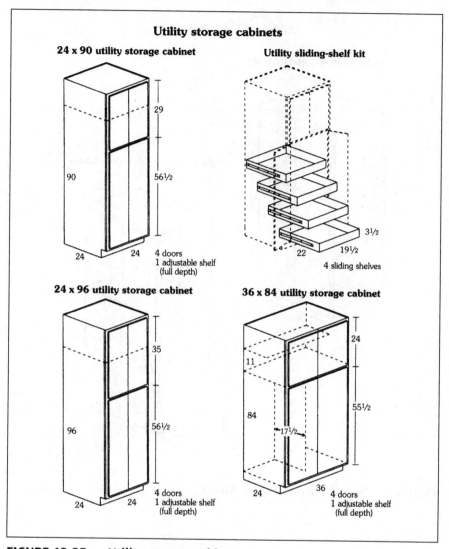

FIGURE 13.25 ▪ **Utility storage cabinets** (Courtesy of Wellborn Cabinet, Inc.)

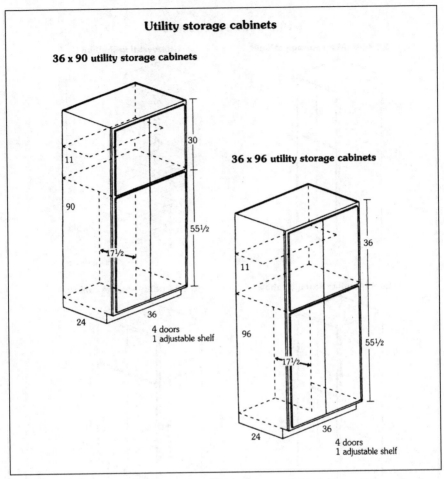

FIGURE 13.26 ■ **Utility storage cabinets** (Courtesy of Wellborn Cabinet, Inc.)

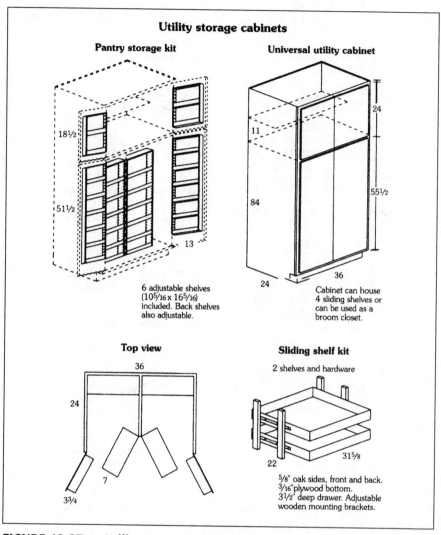

Utility storage cabinets

Pantry storage kit

18½

51½

13

6 adjustable shelves
(10⁵/16 x 16⁵/16)
included. Back shelves
also adjustable.

Universal utility cabinet

24

11

55½

84

24

36

Cabinet can house
4 sliding shelves or
can be used as a
broom closet.

Top view

36

24

7

3³/4

Sliding shelf kit

2 shelves and hardware

31⁵/8

22

⁵/8" oak sides, front and back.
³/16" plywood bottom.
3½" deep drawer. Adjustable
wooden mounting brackets.

FIGURE 13.27 ▪ **Utility storage cabinets** (Courtesy of Wellborn Cabinet, Inc.)

Base accessories

Cutting board

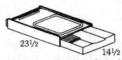

23½
14½

Includes cutting board, knife divider and drawer. Cutting board will replace existing drawer. Use existing drawer front with two hinges that are supplied. Nesessary harware is included also. Cutting board kit is made of oak wood.

Double tiered cutlery/cutting board

Includes cutting board and cutlery divider. Furnished with two hinges and necessary screws for installation. White cutting board is made of nonskid, nonabsorbant polystyrene and is removable. Dishwasher safe. Side flanges may be trimmed.

Interior drawer		
Min. height	Width	Min. depth
3½"	18¼" to 20¾"	17¼"
3½"	12¼" to 14¾"	17¼"

Drawer spice rack

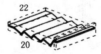

22
20

White high-density polyethylene. May be trimmed.

Silverware tray

Double-tiered white tray.
Guides have expoxy-coated sides.

Top drawer integrates with bottom drawer by sliding back to reveal bottom drawer. Side flanges may be trimmed. Back of drawer must be removed for tray to be functional.

Interior drawer		
Min. height	Width	Min. depth
3½"	12¼" to 14¾"	17¼"

Ironing board

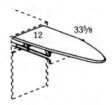

12
33⅝

Built-in ironing board fits into drawer cavity and conveniently pulls out and unfolds.

FIGURE 13.28 ■ **Base cabinet accessories** (Courtesy of Wellborn Cabinet, Inc.)

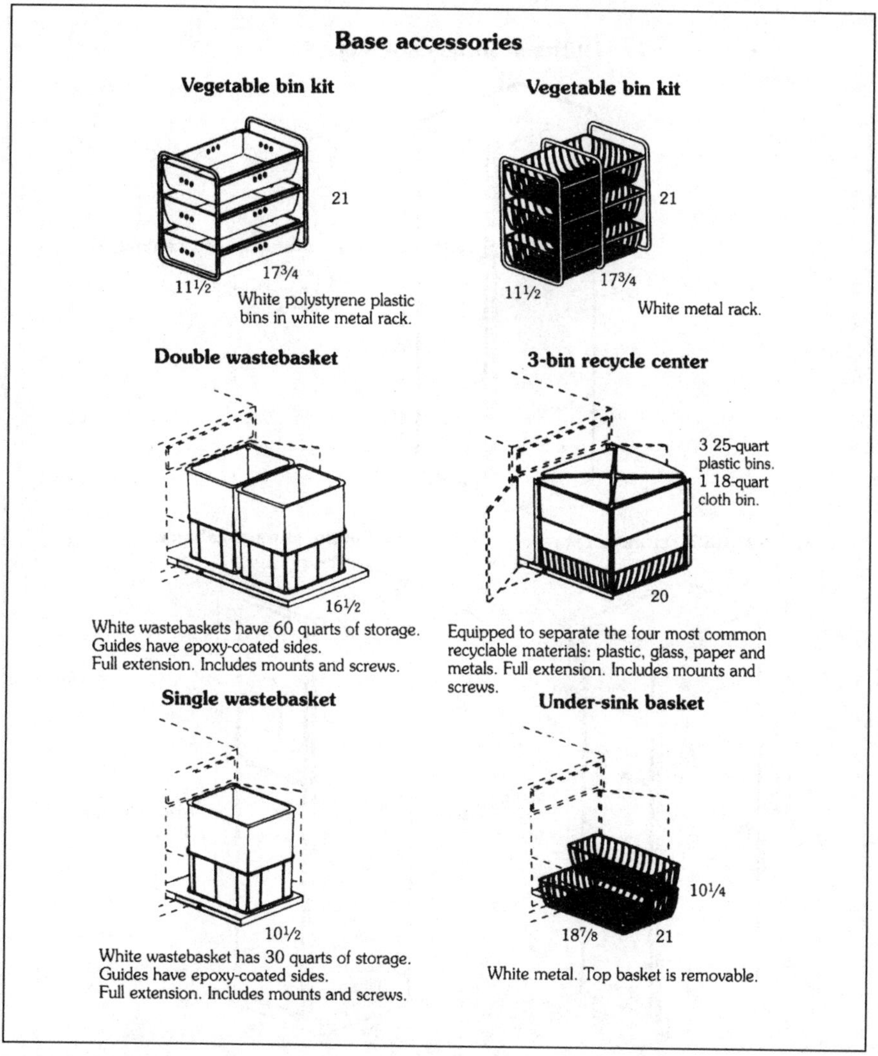

Base accessories

Vegetable bin kit

21

17¾

11½

White polystyrene plastic
bins in white metal rack.

Vegetable bin kit

21

17¾

11½

White metal rack.

Double wastebasket

16½

White wastebaskets have 60 quarts of storage.
Guides have epoxy-coated sides.
Full extension. Includes mounts and screws.

3-bin recycle center

3 25-quart
plastic bins.
1 18-quart
cloth bin.

20

Equipped to separate the four most common
recyclable materials: plastic, glass, paper and
metals. Full extension. Includes mounts and
screws.

Single wastebasket

10½

White wastebasket has 30 quarts of storage.
Guides have epoxy-coated sides.
Full extension. Includes mounts and screws.

Under-sink basket

10¼

18⅞   21

White metal. Top basket is removable.

FIGURE 13.29 ▪ **Base cabinet accessories** (Courtesy of Wellborn Cabinet, Inc.)

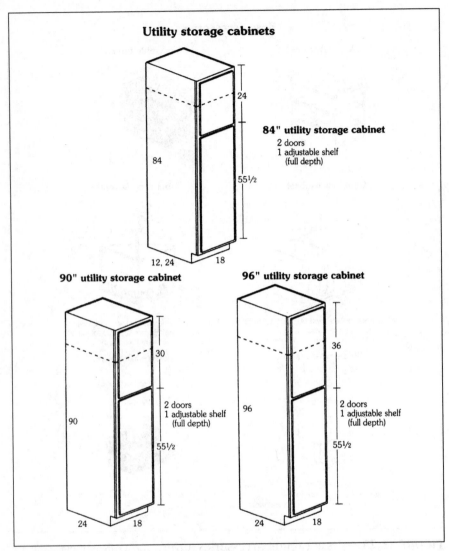

FIGURE 13.30 ▪ **Utility storage cabinets** (Courtesy of Wellborn Cabinet, Inc.)

Base accessories

Silverware tray
1/4" oak.

Sliding shelf kit
5/8" oak sides,
front and back.
3/16" hardwood bottom.
Adjustable wooden
mounting brackets.

17   9,12

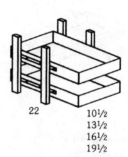

22   10½
          13½
          16½
          19½

Silverware divider
1/4" oak.
Can be trimmed
to 15".

Single roll-out shelf
One shelf and guides
for installation.
1/2" wood grain
laminated
particleboard sides.
5/8" oak front.
Bottom is 1/8"
hardboard.

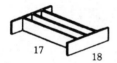

17   18

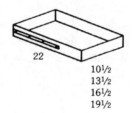

22   10½
          13½
          16½
          19½

FIGURE 13.31 ■ Accessories for base cabinets
(Courtesy of Wellborn Cabinet, Inc.)

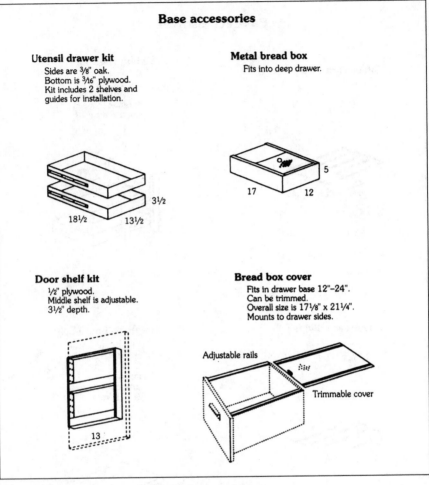

FIGURE 13.32 ▪ Accessories for base cabinets
(Courtesy of Wellborn Cabinet, Inc.)

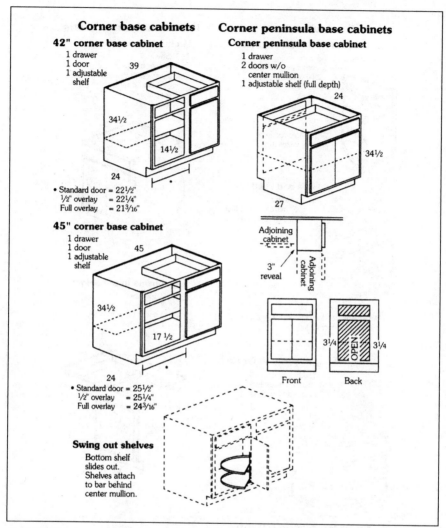

FIGURE 13.33 ■ **Corner base cabinets** (Courtesy of Wellborn Cabinet, Inc.)

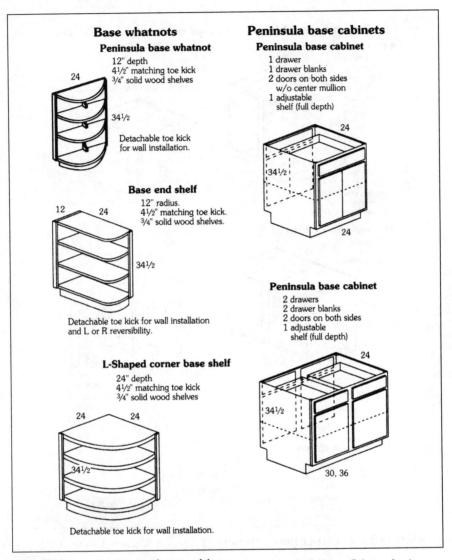

Base whatnots

Peninsula base whatnot

12" depth
4½" matching toe kick
¾" solid wood shelves

24

34½

Detachable toe kick
for wall installation.

Base end shelf

12   24

12" radius.
4½" matching toe kick.
¾" solid wood shelves.

34½

Detachable toe kick for wall installation
and L or R reversibility.

L-Shaped corner base shelf

24" depth
4½" matching toe kick
¾" solid wood shelves

24   24

34½

Detachable toe kick for wall installation.

Peninsula base cabinets

Peninsula base cabinet

1 drawer
1 drawer blanks
2 doors on both sides
  w/o center mullion
1 adjustable
  shelf (full depth)

24

34½

24

Peninsula base cabinet

2 drawers
2 drawer blanks
2 doors on both sides
1 adjustable
  shelf (full depth)

24

34½

30, 36

FIGURE 13.34 ▪ **Corner base cabinets** (Courtesy of Wellborn Cabinet, Inc.)

chapter **14**

CONVERSION TABLES

Conversion tables are very helpful in many situations. Whether you are converting feet to meters or square inches to square feet, a good conversion table can make your life much easier. This chapter is full of useful tables that can speed up your work on a job.

Take some time to review the following tables and become familiar with them. Mark the ones that you feel are most useful in your field of work. While this book is written for carpenters, not all carpenters do the same type of work. So, pick out the tables that mean the most to you and mark them for future reference.

To change	To	Multiply by
Inches	Feet	0.0833
Inches	Millimeters	25.4
Feet	Inches	12
Feet	Yards	0.3333
Yards	Feet	3
Square inches	Square feet	0.00694
Square feet	Square inches	144
Square feet	Square yards	0.11111
Square yards	Square feet	9
Cubic inches	Cubic feet	0.00058
Cubic feet	Cubic inches	1728
Cubic feet	Cubic yards	0.03703
Gallons	Cubic inches	231
Gallons	Cubic feet	0.1337
Gallons	Pounds of water	8.33
Pounds of water	Gallons	0.12004
Ounces	Pounds	0.0625
Pounds	Ounces	16
Inches of water	Pounds per square inch	0.0361
Inches of water	Inches of mercury	0.0735
Inches of water	Ounces per square inch	0.578
Inches of water	Pounds per square foot	5.2
Inches of mercury	Inches of water	13.6
Inches of mercury	Feet of water	1.1333
Inches of mercury	Pounds per square inch	0.4914
Ounces per square inch	Inches of mercury	0.127
Ounces per square inch	Inches of water	1.733
Pounds per square inch	Inches of water	27.72
Pounds per square inch	Feet of water	2.310
Pounds per square inch	Inches of mercury	2.04
Pounds per square inch	Atmospheres	0.0681
Feet of water	Pounds per square inch	0.434
Feet of water	Pounds per square foot	62.5
Feet of water	Inches of mercury	0.8824
Atmospheres	Pounds per square inch	14.696
Atmospheres	Inches of mercury	29.92
Atmospheres	Feet of water	34
Long tons	Pounds	2240
Short tons	Pounds	2000
Short tons	Long tons	0.89295

FIGURE 14.1 ■ Measurement conversion factors

To find	Multiply	By
Microns	Mils	25.4
Centimeters	Inches	2.54
Meters	Feet	0.3048
Meters	Yards	0.19144
Kilometers	Miles	1.609344
Grams	Ounces	28.349523
Kilograms	Pounds	0.4539237
Liters	Gallons (U.S.)	3.7854118
Liters	Gallons (Imperial)	4.546090
Milliliters (cc)	Fluid ounces	29.573530
Milliliters (cc)	Cubic inches	16.387064
Square centimeters	Square inches	6.4516
Square meters	Square feet	0.09290304
Square meters	Square yards	0.83612736
Cubic meters	Cubic feet	2.8316847×10^{-2}
Cubic meters	Cubic yards	0.76455486
Joules	BTU	1054.3504
Joules	Foot-pounds	1.35582
Kilowatts	BTU per minute	0.01757251
Kilowatts	Foot-pounds per minute	2.2597×10^{-5}
Kilowatts	Horsepower	0.7457
Radians	Degrees	0.017453293
Watts	BTU per minute	17.5725

FIGURE 14.2 ▪ Conversion factors in converting from customary (U.S.) units to metric units

	Imperial	Metric
Length	1 inch	25.4 mm
	1 foot	0.3048 m
	1 yard	0.9144 m
	1 mile	1.609 km
Mass	1 pound	0.454 kg
	1 U.S. short ton	0.9072 tonne
Area	1 ft^2	0.092 m^2
	1 yd^2	0.836 m^2
	1 acre	0.404 hectare (ha)
Capacity/Volume	1 ft^3	0.028 m^3
	1 yd^3	0.764 m^3
	1 liquid quart	0.946 litre (1)
	1 gallon	3.785 litre (1)
Heat	1 BTU	1055 joule (J)
	1 BTU/hr	0.293 watt (W)

FIGURE 14.3 ■ Measurement conversions: Imperial to metric

Volume	Weight
1 cu. ft. sand	Approx. 100 lbs.
1 cu. yd.	2700 lbs.
1 ton	¾ yd. or 20 cu. ft.
Avg. shovelful	15 lbs.
12 qt. pail	40 lbs.

FIGURE 14.4 ■ Sand volume to weight conversions

U.S.	Metric
0.001 inch	0.025 mm
1 inch	25.400 mm
1 foot	30.48 cm
1 foot	0.3048 m
1 yard	0.9144 m
1 mile	1.609 km
1 inch^2	6.4516 cm^2
1 feet^2	0.0929 m^2
1 yard^2	0.8361 m^2
1 acre	0.4047 ha
1 mile^2	2.590 km^2
1 inch^3	16.387 cm^3
1 feet^3	0.0283 m^3
1 yard^3	0.7647 m^3
1 U.S. ounce	29.57 ml
1 U.S. pint	0.4732 l
1 U.S. gallon	3.785 l
1 ounce	28.35 g
1 pound	0.4536 kg

FIGURE 14.5 ▪ Conversion tables

Unit	Equals
1 meter	39.3 inches 3.28083 feet 1.0936 yards
1 centimeter	.3937 inch
1 millimeter	.03937 inch, or nearly ⅖ inch
1 kilometer	0.62137 mile
1 foot	.3048 meter
1 inch	2.54 centimeters 25.40 millimeters
1 square meter	10.764 square feet 1.196 square yards
1 square centimeter	.155 square inch
1 square millimeter	.00155 square inch
1 square yard	.836 square meter
1 square foot	.0929 square meter
1 square inch	6.452 square centimeter 645.2 square millimeter

FIGURE 14.6 ▪ Metric-customary equivalents

Unit	Equals
1 cubic meter	35.314 cubic feet 1.308 cubic yards 264.2 U.S. gallons (231 cubic inches)
1 cubic decimeter	61.0230 cubic inches .0353 cubic feet
1 cubic centimeter	.061 cubuic inch
1 liter	1 cubic decimeter 61.0230 cubic inches 0.0353 cubic foot 1.0567 quarts (U.S.) 0.2642 gallon (U.S.) 2.2020 lb. of water at 62°F.
1 cubic yard	.7645 cubic meter
1 cubic foot	.02832 cubic meter 28.317 cubic decimeters 28.317 liters
1 cubic inch	16.383 cubic centimeters
1 gallon (British)	4.543 liters
1 gallon (U.S.)	3.785 liters
1 gram	15.432 grains
1 kilogram	2.2046 pounds
1 metric ton	.9842 ton of 2240 pounds 19.68 cwts. 2204.6 pounds
1 grain	.0648 gram
1 ounce avoirdupois	28.35 grams
1 pound	.4536 kilograms
1 ton of 2240 lb.	1.1016 metric tons 1016 kilograms

FIGURE 14.7 ■ Measures of volume and capacity

Quantity	Unit	Symbol
Time	Second	s
Plane angle	Radius	rad
Force	Newton	N
Energy, work, quantity of heat	Joule	J
	Kilojoule	kJ
	Megajoule	MJ
Power, heat flow rate	Watt	W
	Kilowatt	kW
Pressure	Pascal	Pa
	Kilopascal	kPa
	Megapascal	MPa
Velocity, speed	Meter per second	m/s
	Kilometer per hour	km/h

FIGURE 14.8 ■ Metric symbols

Inches^2	Millimeters^2
0.01227	8.0
0.04909	31.7
0.11045	71.3
0.19635	126.7
0.44179	285.0
0.7854	506.7
1.2272	791.7
1.7671	1140.1
3.1416	2026.8
4.9087	3166.9
7.0686	4560.4
12.566	8107.1
19.635	12667.7
28.274	18241.3
38.485	24828.9
50.265	32478.9
63.617	41043.1
78.540	50670.9

FIGURE 14.9 ■ Area in inches and millimeters

Inches	Millimeters
0.3927	10
0.7854	20
1.1781	30
1.5708	40
2.3562	60
3.1416	80
3.9270	100
4.7124	120
6.2832	160
7.8540	200
9.4248	240
12.566	320
15.708	400
18.850	480
21.991	560
25.133	640
28.274	720
31.416	800

FIGURE 14.10 ▪ Circumference in inches and millimeters

Inches^2	Millimeters^2
0.01227	8.0
0.04909	31.7
0.11045	71.3
0.19635	126.7
0.44179	285.0
0.7854	506.7
1.2272	791.7
1.7671	1140.1
3.1416	2026.8
4.9087	3166.9
7.0686	4560.4
12.566	8107.1
19.635	12667.7
28.274	18241.3
38.485	24828.9
50.265	32478.9
63.617	41043.1
78.540	50670.9

FIGURE 14.11 ▪ Area in inches and millimeters

Feet	Meters (m)	Millimeters (mm)
1	0.305	304.8
2	0.610	609.6
3 (1 yd.)	0.914	914.4
4	1.219	1 219.2
5	1.524	1 524.0
6 (2 yd.)	1.829	1 828.8
7	2.134	2 133.6
8	2.438	2 438.2
9 (3yd.)	2.743	2 743.2
10	3.048	3 048.0
20	6.096	6 096.0
30 (10 yd.)	9.144	9 144.0
40	12.19	12 192.0
50	15.24	15 240.0
60 (20 yd.)	18.29	18 288.0
70	21.34	21 336.0
80	24.38	24 384.0
90 (30 yd.)	27.43	27 432.0
100	30.48	30 480.0

FIGURE 14.12 ■ Length conversions

Units	Equals
1 decimeter	4 inches
1 meter	1.1 yards
1 kilometer	⅝ mile
1 hektar	2½ acres
1 stere or cu. meter	¼ cord
1 liter	1.06 qt. liquid; 0.9 qt. dry
1 hektoliter	2⅝ bushel
1 kilogram	2⅕ lbs.
1 metric ton	2200 lbs.

FIGURE 14.13 ■ Approximate metric equivalents

Inches	Millimeters
1	25.4
2	50.8
3	76.2
4	101.6
5	127.0
6	152.4
7	177.8
8	203.2
9	228.6
10	254.0
11	279.4
12	304.8
13	330.2
14	355.6
15	381.0
16	406.4
17	431.8
18	457.2
19	482.6
20	508.0

FIGURE 14.14 ■ Inches to millimeters

Metric linear measure		
Measure	**Equals**	**Equals**
	1 millimeter	.001 meter
10 millimeter	1 centimeter	.01 meter
10 centimeter	1 decimeter	.1 meter
10 decimeter	1 meter	1 meter
10 meters	1 dekameter	10 meters
10 dekameters	1 hectometer	100 meters
10 hectometers	1 kilometer	1000 meters
10 kilometers	1 myriameter	10,000 meters
Metric land measure		
Unit	**Equals**	
1 centiare (ca.)	1 sq. meter	
100 centiares (ca.)	1 are	
100 ares (A.)	1 hectare	
100 hectares (ha.)	1 sq. kilometer	

FIGURE 14.15 ■ Metric linear measurements

Inches	Meters (m)	Millimeters (mm)
⅛	0.003	3.17
¼	0.006	6.35
⅜	0.010	9.52
½	0.013	12.6
⅝	0.016	15.87
¾	0.019	19.05
⅞	0.022	22.22
1	0.025	25.39
2	0.051	50.79
3	0.076	76.20
4	0.102	101.6
5	0.127	126.9
6	0.152	152.4
7	0.178	177.8
8	0.203	203.1
9	0.229	228.6
10	0.254	253.9
11	0.279	279.3
12	0.305	304.8

FIGURE 14.16 ▪ Length conversions

Quantity	Unit	Symbol
Length	Millimeter	mm
	Centimeter	cm
	Meter	m
	Kilometer	km
Area	Square Millimeter	mm^7
	Square Centimeter	cm^2
	Square Decimeter	dm^2
	Square Meter	m^2
	Square Kilometer	km^2
Volume	Cubic Centimeter	cm^3
	Cubic Decimeter	dm^3
	Cubic Meter	m^3
Mass	Milligram	mg
	Gram	g
	Kilogram	kg
	Tonne	t
Temperature	Degree Celsius	°C
	Kelvin	K
Time	Second	s
Plane angle	Radius	rad
Force	Newton	N
Energy, work, quantity of heat	Joule	J
	Kilojoule	kJ
	Megajoule	MJ
Power, heat flow rate	Watt	W
	Kilowatt	kW
Pressure	Pascal	Pa
	Kilopascal	kPa
	Megapascal	MPa
Velocity, speed	Meter per second	m/s
	Kilometer per hour	km/h
Revolutional frequency	Revolution per minute	r/min

FIGURE 14.17 ▪ Metric symbols

1 cu. ft. at 50°F. weighs 62.41 lb.
1 gal. at 50°F weighs 8.34 lb.
1 cu. ft. of ice weighs 57.2 lb.
Water is at its greatest density at 39.2°F.
1 cu. ft. at 39.2°F. weighs 62.43 lb.

FIGURE 14.18 ▪ Water weight

Quantity	Equals
10 milligrams (mg.)	1 centigram
10 centigrams (cg.)	1 decigram
10 decigrams (dg.)	1 gram
10 grams (g.)	1 dekagram
10 dekagrams (Dg.)	1 hectogram
10 hectograms (hg.)	1 kilogram
10 kilogram (kg.)	1 myriagram
10 myriagrams (Mg.)	1 quintal

FIGURE 14.19 ■ Metric weight measure

Square feet	Square meters
1	0.925
2	.1850
3	.2775
4	.3700
5	.4650
6	.5550
7	.6475
8	.7400
9	.8325
10	.9250
25	2.315
50	4.65
100	9.25

FIGURE 14.20 ■ Square feet to square meters

Quantity	Equals
Metric cubic measure	
1000 cubic millimeters (cu. mm.)	1 cubic centimeter
1000 cubic centimeters (cu. cm.)	1 cubic decimeter
1000 cubic decimeters (cu. dm.)	1 cubic meter
Metric capacity measure	
10 milliliters (mi.)	1 centiliter
10 centiliters (cl.)	1 deciliter
10 deciliters (dl.)	1 liter
10 liters (l.)	1 dekaliter
10 dekaliters (Dl.)	1 hectoliter
10 hectoliters (hl.)	1 kiloliter
10 kiloliters (kl.)	1 myrialiter (ml.)

FIGURE 14.21 ■ Metric cubic measure

To change	To	Multiply by
Inches	Millimeters	25.4
Feet	Meters	.3048
Miles	Kilometers	1.6093
Square inches	Square centimeters	6.4515
Square feet	Square meters	.09290
Acres	Hectares	.4047
Acres	Square kilometers	.00405
Cubic inches	Cubic centimeters	16.3872
Cubic feet	Cubic meters	.02832
Cubic yards	Cubic meters	.76452
Cubic inches	Liters	.01639
U.S. gallons	Liters	3.7854
Ounces (avoirdupois)	Grams	28.35
Pounds	Kilograms	.4536
Lbs. per sq. in. (P.S.I.)	Kg.'s per sq. cm.	.0703
Lbs. per cu. ft.	Kg.'s per cu. meter	16.0189
Tons (2000 lbs.)	Metric tons (1000 kg.)	.9072
Horsepower	Kilowatts	.746

FIGURE 14.22 ■ English to metric conversions

Quantity	Equals
100 sq. millimeters	1 sq. centimeter
100 sq. centimeters	1 sq. decimeter
100 sq. decimeters	1 sq. meter

FIGURE 14.23 ■ Metric square measure

Quantity	Equals
1 meter	39.3 inches 3.28083 feet 1.0936 yards
1 centimeter	.3937 inch
1 millimeter	.03937 inch, or nearly ½₅ inch
1 kilometer	0.62137 mile
.3048 meter	1 foot
2.54 centimeters	1 inch 25.40 millimeters

FIGURE 14.24 ■ Metric conversion

Quantity	Equals	Equals
12 inches	1 foot	
3 feet	1 yard	36 inches
5½ yards	1 rod	16½ feet
40 rods	1 furlong	660 feet
8 furlongs	1 mile	5280 feet

FIGURE 14.25 ■ Linear measure

Quantity	Equals
Linear measure	
12 inches	1 foot
3 feet	1 yard
5½ yards	1 rod
320 rods	1 mile
1 mile	1760 yards
1 mile	5280 feet
Square measure	
144 sq. inches	1 sq. foot
9 sq. feet	1 sq. yard
1 sq. yard	1296 sq. inches
4840 sq. yards	1 acre
640 acres	1 sq. mile
Cubic measure	
1728 cubic inches	1 cubic foot
27 cubic feet	1 cubic yard
Avoirdupois weight	
16 ounces	1 pound
100 pounds	1 hundredweight
20 hundredweight	1 ton
1 ton	2000 pounds
1 long ton	2240

FIGURE 14.26 ■ Weights and measures

Quantity	Equals
4 gills	1 pint
2 pints	1 quart
4 quarts	1 gallon
31½ gallons	1 barrel
1 gallon	231 cubic inches
7.48 gallons	1 cubic foot
1 gallon water	8.33 pounds
1 gallon gasoline	5.84 pounds

FIGURE 14.27 ■ Liquid measure

Unit	Equals
1 cu. ft.	62.4 lbs.
1 cu. ft.	7.48 gal.
1 gal.	8.33 lbs.
1 gal.	0.1337 cu. ft.

FIGURE 14.28 ▪ Water volume to weight conversion

Unit	Equals
1 gallon	0.133681 cubic foot
1 gallon	231 cubic inches

FIGURE 14.29 ▪ Volume measure equivalents

Quantity	Equals
12 inches (in. or ")	1 foot (ft. or ')
3 feet	1 yard (yd.)
5½ yards or 16½ feet	1 rod (rd.)
40 rods	1 furlong (fur.)
8 furlongs or 320 rods	1 statute mile (mi.)

FIGURE 14.30 ▪ Long measure

Unit	Equals
1 sq. centimeter	0.1550 sq. in.
1 sq. decimeter	0.1076 sq. ft.
1 sq. meter	1.196 sq. yd.
1 are	3.954 sq. rd.
1 hektar	2.47 acres
1 sq. kilometer	0.386 sq. mile
1 sq. in.	6.452 sq. centimeters
1 sq. ft.	9.2903 sq. decimeters
1 sq. yd.	0.8361 sq. meter
1 sq. rd.	0.2529 are
1 acre	0.4047 hektar
1 sq. mile	2.59 sq. kilometers

FIGURE 14.31 ▪ Square measure

Quantity	Equals
1 square meter	10.764 square feet 1.196 square yards
1 square centimeter	.155 square inch
1 square millimeter	.00155 square inch
.836 square meter	1 square yard
.0929 square meter	1 square foot
6.452 square centimeter	1 square inch
645.2 square millimeter	1 square inch

FIGURE 14.32 ■ Surface measures

Quantity	Equals	Equals
7.92 inches	1 link	
100 links	1 chain	66 feet
10 chains	1 furling	660 feet
80 chains	1 mile	5280 feet

FIGURE 14.33 ■ Surveyor's measure

Quantity	Equals
144 sq. inches	1 sq. foot
9 sq. feet	1 sq. yard
1 sq. yard	1296 sq. inches
4840 sq. yards	1 acre
640 acres	1 sq. mile

FIGURE 14.34 ■ Square measure

Inches	Decimals of an inch
¹⁄₆₄	.0156
¹⁄₃₂	.0312
³⁄₆₄	.0468
¹⁄₁₆	.0625
⁵⁄₆₄	.0781
³⁄₃₂	.0937
⁷⁄₆₄	.1093
¹⁄₈	.1250
⁹⁄₆₄	.1406
⁵⁄₃₂	.1562
¹¹⁄₆₄	.1718
³⁄₁₆	.1875
¹³⁄₆₄	.2031
⁷⁄₃₂	.2187
¹⁵⁄₆₄	.2343
¹⁄₄	.2500
¹⁷⁄₆₄	.2656
⁹⁄₃₂	.2812
¹⁹⁄₆₄	.2968
⁵⁄₁₆	.3125

FIGURE 14.35 ■ Decimal equivalents of fractions

Inches	Millimeters
¹⁄₈	3.2
¹⁄₄	6.4
³⁄₈	9.5
¹⁄₂	12.7
³⁄₄	19.1
1	25.4
1¹⁄₄	31.8
1¹⁄₂	38.1
2	50.8
2¹⁄₂	63.5
3	76.2
4	101.6
5	127
6	152.4
7	177.8
8	203.2
9	228.6
10	254

FIGURE 14.36 ■ Diameter in inches and millimeters

Unit	Equals
1 gram	15.432 grains
1 kilogram	2.2046 pounds
1 metric ton	.9842 ton of 2240 pounds
	19.68 cwts.
	2204.6 pounds
1 grain	.0648 gram
1 ounce avoirdupois	28.35 grams
1 pound	.4536 kilograms
1 ton of 2240 lb.	1.1016 metric tons
	1016 kilograms

FIGURE 14.37 ■ Weight conversions

Quantity	Equals	Meters	English equivalent
1 mm.	1 millimeter	1/1000	.03937 in.
10 mm.	1 centimeter	1/100	.3937 in.
10 cm.	1 decimeter	1/10	3.937 in.
10 dm.	1 meter	1	39.37 in.
10 m.	1 dekameter	10	32.8 ft.
10 Dm.	1hectometer	100	328.09 ft.
10 Hm.	1 kilometer	1000	.62137 mile

FIGURE 14.38 ■ Lengths

Quantity	Equals	Equals
144 sq. inches	1 sq. foot	
9 sq. feeet	1 sq. yard	
30¼ sq. yards	1 sq. rod	272.25 sq. feet
160 sq. rods	1 acre	4840 sq. yards
		or 43,560 sq. feet
640 acres	1 sq. mile	3,097,600 sq. yards
36 sq. miles	1 township	

FIGURE 14.39 ■ Square measure

Quantity	Equals	Cubic inches
2 pints	1 quart	67.2
8 quarts	1 peck	537.61
4 pecks	1 bushel	2150.42

FIGURE 14.40 ■ Dry measure

Quantity	Equals
144 sq. in.	1 sq. ft.
9 sq. ft.	1 sq. yd.
30½ sq. yd.	1 sq. rd.
160 sq. rd.	1 acre
640 acres	1 sq. mile
43,560 sq. ft.	1 acre

FIGURE 14.41 ■ Surface measure

Passageway	Recommended	Minimum
Stairs	40"	36"
Landings	40"	36"
Main hall	48"	36"
Minor hall	36"	30"
Interior door	32"	28"
Exterior door	36"	36"

FIGURE 14.42 ■ Widths of passageways

Barometer (ins. of mercury)	Pressure (lbs. per sq. in.)
28.00	13.75
28.25	13.88
28.50	14.00
28.75	14.12
29.00	14.24
29.25	14.37
29.50	14.49
29.75	14.61
29.921	14.696
30.00	14.74
30.25	14.86
30.50	14.98
30.75	15.10
31.00	15.23

Rule: Barometer in inches of mercury × .49116 = lbs. per sq. in.

FIGURE 14.43 ■ Atmospheric pressure per square inch

To change	To	Multiply by
Inches	Feet	0.0833
Inches	Millimeters	25.4
Feet	Inches	12
Feet	Yards	0.3333
Yards	Feet	3
Square inches	Square feet	0.00694
Square feet	Square inches	144
Square feet	Square yards	0.11111
Square yards	Square feet	9
Cubic inches	Cubic feet	0.00058
Cubic feet	Cubic inches	1728
Cubic feet	Cubic yards	0.03703
Cubic yards	Cubic feet	27
Cubic inches	Gallons	0.00433
Cubic feet	Gallons	7.48
Gallons	Cubic inches	231
Gallons	Cubic feet	0.1337
Gallons	Pounds of water	8.33
Pounds of water	Gallons	0.12004
Ounces	Pounds	0.0625
Pounds	Ounces	16
Inches of water	Pounds per square inch	0.0361
Inches of water	Inches of mercury	0.0735
Inches of water	Ounces per square inch	0.578
Inches of water	Pounds per square foot	5.2
Inches of mercury	Inches of water	13.6
Inches of mercury	Feet of water	1.1333
Inches of mercury	Feet of water	0.4914
Ounces per square inch	Pounds per square inch	0.127
Ounces per square inch	Inches of mercury	1.733
Pounds per square inch	Inches of water	27.72
Pounds per square inch	Feet of water	2.310
Pounds per square inch	Inches of mercury	2.04
Pounds per square inch	Atmospheres	0.0681

FIGURE 14.44 ▪ Useful multipliers

chapter **15**

MATH FOR THE TRADES

The illustrations in this chapter will help you to understand math for the trades. You will find formulas, conversion tables, and reference tables. All of these tools are designed to make your life easier.

Some of what you will see might seem strange. You may wonder why you need to know how to find the volume of a cylinder. Have you ever poured concrete in a pier foundation for a deck? If so, you had to know how much concrete would be needed to fill the round hole. Once you think about some of the illustrations, you will see how they fit into your life in the field.

☑ *fast***facts**

Carpenters do a lot of math during the course of a day. Most of them never think about it. Why? Because it is second nature to them and an important element of their trade. If you asked the average carpenter to explain trigonometry or algebra, you would probably be faced with a scowling face. The fact is, many carpenters do all sorts of math and simply think of it as part of the job. The illustrations in this chapter will help you to make the most of your math skills

Function	Formula
Sine	$\sin = \dfrac{\text{side opposite}}{\text{hypotenuse}}$
Cosine	$\cos = \dfrac{\text{side adjacent}}{\text{hypotenuse}}$
Tangent	$\tan = \dfrac{\text{side opposite}}{\text{side adjacent}}$
Cosecant	$\csc = \dfrac{\text{hypotenuse}}{\text{side opposite}}$
Secant	$\sec = \dfrac{\text{hypotenuse}}{\text{side adjacent}}$
Cotangent	$\cot = \dfrac{\text{side adjacent}}{\text{side opposite}}$

FIGURE 15.1 ■ Trigonometry

Multiply Length × Width × Thickness
Example: 50 ft. × 10 ft. × 8 in.
50' × 10' × 8/12' = 333.33 cu. feet
To convert to cubic yards, divide by 27 cu. ft. per cu. yd.
Example: 333.33 ÷ 27 = 12.35 cu. yd.

FIGURE 15.2 ■ Estimating volume

Area of surface = Diameter × 3.1416 × length + area of the two bases
Area of base = Diameter × diameter × .7854
Area of base = Volume ÷ length
Length = Volume ÷ area of base
Volume = Length × area of base
Capacity in gallons = Volume in inches ÷ 231
Capacity of gallons = Diameter × diameter × length × .0034
Capacity in gallons = Volume in feet × 7.48

FIGURE 15.3 ■ Cylinder formulas

Area = Short diameter × long diameter × .7854

FIGURE 15.4 ■ Ellipse calculation

Area of surface = One half of circumference of base × slant height + area
of base.
Volume = Diameter × diameter × .7854 × one-third of the altitude.

FIGURE 15.5 ∎ **Cone calculation**

Volume = Width × height × length

FIGURE 15.6 ∎ **Volume of a rectangular prism**

Area = Length × width

FIGURE 15.7 ∎ **Finding the area of a square**

Area = ½ perimeter of base × slant height + area of base
Volume = Area of base × ⅓ of the altitude

FIGURE 15.8 ∎ **Finding area and volume of a pyramid**

These comprise the numerous figures having more than four sides, names
according to the number of sides, thus:

Figure	Sides
Pentagon	5
Hexagon	6
Heptagon	7
Octagon	8
Nonagon	9
Decagon	10

To find the area of a polygon: Multiply the sum of the sides (perimeter of
the polygon) by the perpendicular dropped from its center to one of its
sides, and half the product will be the area. This rule applies to all regular
polygons.

FIGURE 15.9 ∎ **Polygons**

Area = Width of side × 2.598 × width of side

FIGURE 15.10 ∎ **Hexagons**

Area = Base × distance between the two parallel sides

FIGURE 15.11 ■ Parallelograms

Area = Length × width

FIGURE 15.12 ■ Rectangles

Area of surface = Diameter × diameter × 3.1416
Side of inscribed cube = Radius × 1.547
Volume = Diameter × diameter × diameter × .5236

FIGURE 15.13 ■ Spheres

Area = One-half of height times base

FIGURE 15.14 ■ Triangles

Area = One-half of the sum of the parallel sides × the height

FIGURE 15.15 ■ Trapezoids

Volume = Width × height × length

FIGURE 15.16 ■ Cubes

Circumference = Diameter × 3.1416
Circumference = Radius × 6.2832
Diameter = Radius × 2
Diameter = Square root of (area ÷ .7854)
Diameter = Square root of area × 1.1283
Diameter = Circumference × .31831
Radius = Diameter ÷ 2
Radius = Circumference × .15915
Radius = Square root of area × .56419
Area = Diameter × Diameter × .7854
Area = Half of the circumference × half of the diameter
Area = Square of the circumference × .0796
Arc length = Degrees × radius × .01745
Degrees of arc = Length ÷ (radius × .01745)
Radius of arc = Length ÷ (degrees × .01745)
Side of equal square = Diameter × .8862
Side of inscribed square = Diameter × .7071
Area of sector = Area of circle × degrees of arc ÷ 360

FIGURE 15.17 ▪ Formulas for a circle

1. Circumference of a circle = π × diameter or 3.1416 × diameter
2. Diameter of a circle = Circumference × .31831
3. Area of a square = Length × width
4. Area of a rectangle = Length × width
5. Area of a parallelogram = Base × perpendicular height
6. Area of a triangle = ½ base × perpendicular height
7. Area of a circle = π × radius squared or diameter squared × .7854
8. Area of an ellipse = Length × width × .7854
9. Volume of a cube or rectangular prism = Length × width × height
10. Volue of a triangular prism = Area of triangle × length
11. Volume of a sphere = Diameter cubed × .5236 or (dia. × dia. × dia. × .5236)
12. Volume of a cone = π × radius square × ⅓ height
13. Volume of a cylinder = π × radius squared × height
14. Length of one side of a square × 1.128 = Diameter of an equal circle
15. Doubling the diameter of a pipe or cylinder increases its capacity 4 times
16. The pressure (in lbs. per sq. inch) of a column of water = Height of the column (in feet) × .434
17. The capacity of a pipe or tank (in U.S. gallons) = Diameter squared (in inches) × the length (in inches) × .0034
18. A gallon of water = 8⅓ lb. = 231 cu. inches
19. A cubic foot of water = 62½ lb. = 7½ gallons

FIGURE 15.18 ▪ Useful formulas

MULTIPLY	BY	TO OBTAIN
Gallons/minute	8.0208	Cubic feet/hour
Gallons water/minute	6.0086	Tons of water/24 hours
Inches	2.540	Centimeters
Inches of mercury	0.03342	Atmospheres
Inches of mercury	1.133	Feet of water
Inches of mercury	0.4912	Pounds/square inch
Inches of water	0.002458	Atmospheres
Inches of water	0.07355	Inches of mercury
Inches of water	5.202	Pounds/square feet
Inches of water	0.03613	Pounds/square inch
Liters	1000	Cubic centimeters
Liters	61.02	Cubic inches
Liters	0.2642	Gallons
Miles	5280	Feet
Miles/hour	88	Feet/minute
Miles/hour	1.467	Feet/second
Millimeters	0.1	Centimeters
Millimeters	0.03937	Inches
Million gallon/day	1.54723	Cubic feet/second
Pounds of water	0.01602	Cubic feet
Pounds of water	27.68	Cubic inches
Pounds of water	0.1198	Gallons
Pounds/cubic inch	1728	Pounds/cubic feet
Pounds/square foot	0.01602	Feet of water
Pounds/square inch	0.06804	Atmospheres
Pounds/square inch	2.307	Feet of water
Pounds/square inch	2.036	Inches of mercury
Quarts (dry)	67.20	Cubic inches
Quarts (liquid)	57.75	Cubic inches
Square feet	144	Square inches
Square miles	640	Acres
Square yards	9	Square feet
Temperature (°C) + 273	1	Abs. temperature (°C)
Temperature (°C) + 17.28	1.8	Temperature (°F)
Temperature (°F) + 460	1	Abs. temperature (°F)
Temperature (°F) - 32	5/9	Temperature (°C)
Tons (short)	2000	Pounds
Tons of water/24 hours	83.333	Pounds water/hour
Tons of water/24 hours	0.16643	Gallons/minute
Tons of water/24 hours	1.3349	Cubic feet/hour

FIGURE 15.19 ■ A useful set of tables to keep on hand (Reprinted from the 2000 Uniform Plumbing Code (UPC) with the permission of the International Association of Plumbing and Mechanical Officials (IAPMO))

AREAS AND CIRCUMFERENCE OF CIRCLES

Diameter		Circumference		Area	
Inches	mm	Inches	mm	Inches^2	mm^2
1/8	6	0.40	10	0.01227	8.0
1/4	8	0.79	20	0.04909	31.7
3/8	10	1.18	30	0.11045	71.3
1/2	15	1.57	40	0.19635	126.7
3/4	20	2.36	60	0.44179	285.0
1	25	3.14	80	0.7854	506.7
1-1/4	32	3.93	100	1.2272	791.7
1-1/2	40	4.71	120	1.7671	1140.1
2	50	6.28	160	3.1416	2026.8
2-1/2	65	7.85	200	4.9087	3166.9
3	80	9.43	240	7.0686	4560.4
4	100	12.55	320	12.566	8107.1
5	125	15.71	400	19.635	12,667.7
6	150	18.85	480	28.274	18,241.3
7	175	21.99	560	38.485	24,828.9
8	200	25.13	640	50.265	32,428.9
9	225	28.27	720	63.617	41,043.1
10	250	31.42	800	78.540	50,670.9

EQUAL PERIPHERIES

$S = 0.7854\,D$

$D = 1.2732\,S$

$S = 0.8862\,D$

$D = 1.1284\,S$

$S = 0.2821\,C$

EQUAL AREAS

Area of square (S') = 1.2732 × area of circle

Area of square (S) = 0.6366 × area of circle

$C = \pi D = 2\pi R$

$C = 3.5446\,\sqrt{area}$

$D = 0.3183\,C = 2R$

$D = 1.1283\,\sqrt{area}$

$Area = \pi R^2 = 0.7854\,D^2$

$Area = 0.07958\,C^2 = \dfrac{\pi D^2}{4}$

$\pi = 3.1416$

FIGURE 15.20 ■ **More useful information** (Reprinted from the 2000 Uniform Plumbing Code (UPC) with the permission of the International Association of Plumbing and Mechanical Officials (IAPMO))

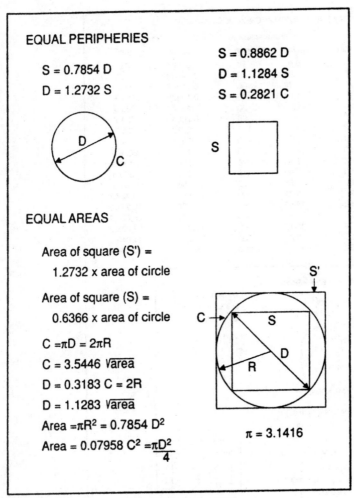

EQUAL PERIPHERIES

S = 0.7854 D

D = 1.2732 S

S = 0.8862 D

D = 1.1284 S

S = 0.2821 C

EQUAL AREAS

Area of square (S') =
   1.2732 x area of circle

Area of square (S) =
   0.6366 x area of circle

$C = \pi D = 2\pi R$

$C = 3.5446 \sqrt{area}$

$D = 0.3183\ C = 2R$

$D = 1.1283 \sqrt{area}$

$Area = \pi R^2 = 0.7854\ D^2$

$Area = 0.07958\ C^2 = \dfrac{\pi D^2}{4}$

$\pi = 3.1416$

FIGURE 15.21 ■ Mathematical formulas

Parallelogram	Area = base × distance between the two parallel sides
Pyramid	Area = ½ perimeter of base × slant height + area of base
	Volume = area of base × ⅓ of the altitude
Rectangle	Area = length × width
Rectangular prism	Volume = width × height × length
Sphere	Area of surface = diameter × diameter × 3.1416
	Side of inscribed cube = radius × 1.547
	Volume = diameter × diameter × diameter × 0.5236
Square	Area = length × width
Triangle	Area = one-half of height times base
Trapezoid	Area = one-half of the sum of the parallel sides × the height
Cone	Area of surface = one-half of circumference of base × slant height + area of base
	Volume = diameter × diameter × 0.7854 × one-third of the altitude
Cube	Volume = width × height × length
Ellipse	Area = short diameter × long diameter × 0.7854
Cylinder	Area of surface = diameter × 3.1416 × length + area of the two bases
	Area of base = diameter × diameter × 0.7854
	Area of base = volume ÷ length
	Length = volume ÷ area of base
	Volume = length × area of base
	Capacity in gallons = volume in inches ÷ 231
	Capacity of gallons = diameter × diameter × length × 0.0034
	Capacity in gallons = volume in feet × 7.48
Circle	Circumference = diameter × 3.1416
	Circumference = radius × 6.2832
	Diameter = radius × 2
	Diameter = square root of = (area ÷ 0.7854)
	Diameter = square root of area × 1.1233

FIGURE 15.22 ■ Area and other formulas

	−100°–30°	
°C	Base temperature	°F
−73	−100	−148
−68	−90	−130
−62	−80	−112
−57	−70	−94
−51	−60	−76
−46	−50	−58
−40	−40	−40
−34.4	−30	−22
−28.9	−20	−4
−23.3	−10	14
−17.8	0	32
−17.2	1	33.8
−16.7	2	35.6
−16.1	3	37.4
−15.6	4	39.2
−15.0	5	41.0
−14.4	6	42.8
−13.9	7	44.6
−13.3	8	46.4
−12.8	9	48.2
−12.2	10	50.0
−11.7	11	51.8
−11.1	12	53.6
−10.6	13	55.4
−10.0	14	57.2
	31°–71°	
°C	Base temperature	°F
−0.6	31	87.8
0	32	89.6
0.6	33	91.4
1.1	34	93.2
1.7	35	95.0
2.2	36	96.8
2.8	37	98.6
3.3	38	100.4
3.9	39	102.2
4.4	40	104.0
5.0	41	105.8
5.6	42	107.6

FIGURE 15.23 ■ Temperature coversion

Vacuum in inches of mercury	Boiling point
29	76.62
28	99.93
27	114.22
26	124.77
25	133.22
24	140.31
23	146.45
22	151.87
21	156.75
20	161.19
19	165.24
18	169.00
17	172.51
16	175.80
15	178.91
14	181.82
13	184.61
12	187.21
11	189.75
10	192.19
9	194.50
8	196.73
7	198.87
6	200.96
5	202.25
4	204.85
3	206.70
2	208.50
1	210.25

FIGURE 15.24 ■ Boiling points of water based on pressure

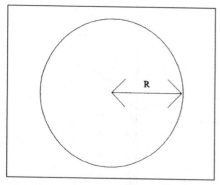

FIGURE 15.25 ■ Radius of a circle

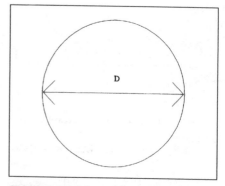

FIGURE 15.26 ■ Diameter of a circle

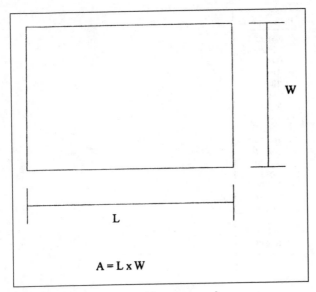

FIGURE 15.27 ■ Area of a rectangle

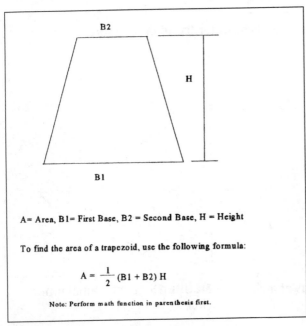

A= Area, B1= First Base, B2 = Second Base, H = Height

To find the area of a trapezoid, use the following formula:

$$A = \frac{1}{2}(B1 + B2)\,H$$

Note: Perform math function in parenthesis first.

FIGURE 15.28 ■ Area of a trapezoid

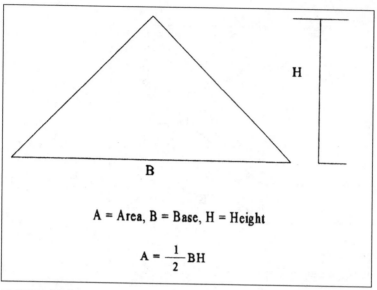

FIGURE 15.29 ■ Area of a triangle

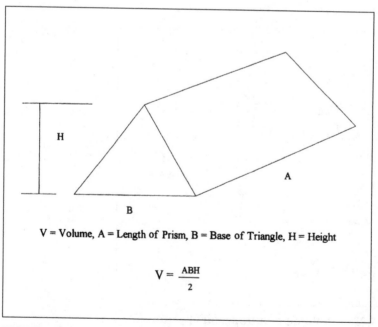

FIGURE 15.30 ■ Area of a triangular prism

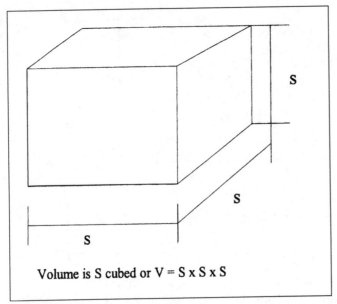

Volume is S cubed or V = S x S x S

FIGURE 15.31 ■ Volume of a cube

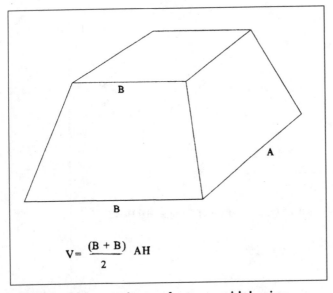

$$V = \frac{(B + B)}{2} AH$$

FIGURE 15.32 ■ Volume of a trapezoidal prism

Set	Travel	Set	Travel	Set	Travel
2	2.828	¼	15.907	½	28.987
¼	3.181	½	16.261	¾	29.340
½	3.531	¾	16.614	21	29.694
¾	3.888	12	16.968	¼	30.047
3	4.242	¼	17.321	½	30.401
¼	4.575	½	17.675	¾	30.754
½	4.949	¾	18.028	22	31.108
¾	5.302	13	18.382	¼	31.461
4	5.656	¼	18.735	½	31.815
¼	6.009	½	19.089	¾	32.168
½	6.363	¾	19.442	23	32.522
¾	6.716	14	19.796	¼	32.875
5	7.070	¼	20.149	½	33.229
¼	7.423	½	20.503	¾	33.582
½	7.777	¾	20.856	24	33.936
¾	8.130	15	21.210	¼	34.289
6	8.484	¼	21.563	½	34.643
¼	8.837	½	21.917	¾	34.996
½	9.191	¾	22.270	25	35.350
¾	9.544	16	22.624	¼	35.703
7	9.898	¼	22.977	½	36.057
¼	10.251	½	23.331	¾	36.410
½	10.605	¾	23.684	26	36.764
¾	10.958	17	24.038	¼	37.117
8	11.312	¼	24.391	½	37.471
¼	11.665	½	24.745	¾	37.824
½	12.019	¾	25.098	27	38.178
¾	12.372	18	25.452	¼	38.531
9	12.726	¼	25.805	½	38.885
¼	13.079	½	26.159	¾	39.238
½	13.433	¾	26.512	28	39.592
¾	13.786	19	26.866	¼	39.945
10	14.140	¼	27.219	½	40.299
¼	14.493	½	27.573	¾	40.652
½	14.847	¾	27.926	29	41.006
¾	15.200	20	28.280	¼	41.359
11	15.554	¼	28.635	½	41.713

FIGURE 15.33 ■ Set and travel relationships in inches for 45° offsets

Inches	Decimal of an inch	Inches	Decimal of an inch
1/64	.015625	33/64	.515625
1/32	.03125	17/32	.53125
3/64	.046875	35/64	.546875
1/16	.0625	9/16	.5625
5/64	.078125	37/64	.578125
3/32	.09375	19/32	.59375
7/64	.109375	39/64	.609375
1/8	.125	5/8	.625
9/64	.140625	41/64	.640625
5/32	.15625	21/32	.65625
11/64	.171875	43/64	.671875
3/16	.1875	11/16	.6875
13/64	.203125	45/64	.703125
7/32	.21875	23/32	.71875
15/64	.234375	47/64	.734375
1/4	.25	3/4	.75
17/64	.265625	49/64	.765625
9/32	.28125	25/32	.78125
19/64	.296875	51/64	.796875
5/16	.3125	13/16	.8125
21/64	.328125	53/64	.828125
11/32	.34375	27/32	.84375
23/64	.359375	55/64	.859375
3/8	.375	7/8	.875
35/64	.390625	57/64	.890625
13/32	.40625	22/32	.90625
27/64	.421875	59/64	.921875
7/16	.4375	15/16	.9375
29/64	.453125	61/64	.953125
15/32	.46875	31/32	.96875
31/64	.484375	63/64	.984375
1/2	.5	1	1

FIGURE 15.34 ■ Simple offsets

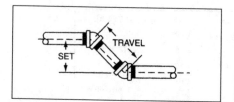

FIGURE 15.35 ■ Calculated 45° offsets

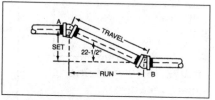

FIGURE 15.36 ■ Decimal equivalents of fractions of an inch

To find side*	When known side is	Multiply Side	For 60° ells by	For 45° ells by	For 30° ells by	For 22½° ells by	For 11¼° ells by	For 5⅝° ells by
T	S	S	1.155	1.414	2.000	2.613	5.125	10.187
S	T	T	.866	.707	.500	.383	.195	.098
R	S	S	.577	1.000	1.732	2.414	5.027	10.158
S	R	R	1.732	1.000	.577	.414	.198	.098
T	R	R	2.000	1.414	1.155	1.082	1.019	1.004
R	T	T	.500	.707	.866	.924	.980	.995

*S = set, R = run, T = travel.

FIGURE 15.37 ■ Multipliers for calculating simple offsets

Metric	U.S.
144 in.²	1 ft²
9 ft²	1 yd²
1 yd²	1296 in.²
4840 yd²	1 a
640 a	1 mi²

FIGURE 15.38 ■ Square measure

1 cm²	0.1550 in.²
1 dm²	0.1076 ft²
1 ms2	1.196 yd²
1 A (are)	3.954 rd²
1 ha	2.47 a (acres)
1 km²	0.386 mi²
1 in.²	6.452 cm²
1 ft²	9.2903 dm²
1 yd²	0.8361 m²
1 rd²	0.2529 A (are)
1 a (acre)	0.4047 ha
1 mi²	2.59 km²

FIGURE 15.39 ■ Square measures of length and area

100 mm²	1 cm²
100 cm²	1 dm²
100 dm²	1 m²

FIGURE 15.40 ■ Metric square measure

144 in.²	1 ft²	272.25 ft²
9 ft²	1 yd²	
30¼ yd²	1 rd²	272.25 ft²
160 rd²	1 a	4840 yd²
		43,560 ft²)
640 a	1 m²	3,097,600 yd²
36 mi²	1 township	

FIGURE 15.41 ■ Square measures

Square feet	Square meters
1	0.925
2	0.1850
3	0.2775
4	0.3700
5	0.4650
6	0.5550
7	0.6475
8	0.7400
9	0.8325
10	0.9250
25	2.315
50	4.65
100	9.25

FIGURE 15.42 ■ Square feet to square meters

Fraction	Square root
1/8	.3535
1/4	.5000
3/8	.6124
1/2	.7071
5/8	.7906
3/4	.8660
7/8	.9354

FIGURE 15.43 ■ Square roots of fractions

Fraction	Cube root
1/8	.5000
1/4	.6300
3/8	.7211
1/2	.7937
5/8	.8550
3/4	.9086
7/8	.9565

FIGURE 15.44 ■ Cube roots of fractions

Number	Cube	Number	Cube	Number	Cube
1	1	36	46,656	71	357,911
2	8	37	50.653	72	373,248
3	27	38	54,872	73	389,017
4	64	39	59,319	74	405,224
5	125	40	64,000	75	421,875
6	216	41	68,921	76	438,976
7	343	42	74,088	77	456,533
8	512	43	79,507	78	474,552
9	729	44	85,184	79	493,039
10	1,000	45	91,125	80	512,000
11	1,331	46	97,336	81	531,441
12	1,728	47	103,823	82	551,368
13	2,197	48	110,592	83	571,787
14	2,477	49	117,649	84	592,704
15	3,375	50	125,000	85	614,125
16	4,096	51	132,651	86	636,056
17	4,913	52	140,608	87	658,503
18	5,832	53	148,877	88	681,472
19	6,859	54	157,464	89	704,969
20	8,000	55	166,375	90	729,000
21	9,621	56	175,616	91	753,571
22	10,648	57	185,193	92	778,688
23	12,167	58	195,112	93	804,357
24	13,824	59	205,379	94	830,584
25	15,625	60	216,000	95	857,375
26	17,576	61	226,981	96	884,736
27	19,683	62	238,328	97	912,673
28	21,952	63	250,047	98	941,192
29	24,389	64	262,144	99	970,299
30	27,000	65	274,625	100	1,000,000
31	29,791	66	287,496		
32	32,768	67	300,763		
33	35,937	68	314,432		
34	39,304	69	328,500		
35	42,875	70	343,000		

FIGURE 15.45 ▪ Cubes of numbers

Number	Square root	Number	Square root	Number	Square root
1	1.00000	36	6.00000	71	8.42614
2	1.41421	37	6.08276	72	8.48528
3	1.73205	38	6.16441	73	8.54400
4	2.00000	39	6.24499	74	8.60232
5	2.23606	40	6.32455	75	8.66025
6	2.44948	41	6.40312	76	8.71779
7	2.64575	42	6.48074	77	8.77496
8	2.82842	43	6.55743	78	8.83176
9	3.00000	44	6.63324	79	8.88819
10	3.16227	45	6.70820	80	8.94427
11	3.31662	46	6.78233	81	9.00000
12	3.46410	47	6.85565	82	9.05538
13	3.60555	48	6.92820	83	9.11043
14	3.74165	49	7.00000	84	9.16515
15	3.87298	50	7.07106	85	9.21954
16	4.00000	51	7.14142	86	9.27361
17	4.12310	52	7.21110	87	9.32737
18	4.24264	53	7.28010	88	9.38083
19	4.35889	54	7.34846	89	9.43398
20	4.47213	55	7.41619	90	9.48683
21	4.58257	56	7.48331	91	9.53939
22	4.69041	57	7.54983	92	9.59166
23	4.79583	58	7.61577	93	9.64365
24	4.89897	59	7.68114	94	9.69535
25	5.00000	60	7.74596	95	9.74679
26	5.09901	61	7.81024	96	9.79795
27	5.19615	62	7.87400	97	9.84885
28	5.29150	63	7.93725	98	9.89949
29	5.38516	64	8.00000	99	9.94987
30	5.47722	65	8.06225	100	10.00000
31	5.56776	66	8.12403		
32	5.65685	67	8.18535		
33	5.74456	68	8.24621		
34	5.83095	69	8.30662		
35	5.91607	70	8.36660		

FIGURE 15.46 ■ Square roots of numbers

Number	Square	Number	Square	Number	Square
1	1	36	1296	71	5041
2	4	37	1369	72	5184
3	9	38	1444	73	5329
4	16	39	1521	74	5476
5	25	40	1600	75	5625
6	36	41	1681	76	5776
7	49	42	1764	77	5929
8	64	43	1849	78	6084
9	81	44	1936	79	6241
10	100	45	2025	80	6400
11	121	46	2116	81	6561
12	144	47	2209	82	6724
13	169	48	2304	83	6889
14	196	49	2401	84	7056
15	225	50	2500	85	7225
16	256	51	2601	86	7396
17	289	52	2704	87	7569
18	324	53	2809	88	7744
19	361	54	2916	89	7921
20	400	55	3025	90	8100
21	441	56	3136	91	8281
22	484	57	3249	92	8464
23	529	58	3364	93	8649
24	576	59	3481	94	8836
25	625	60	3600	95	9025
26	676	61	3721	96	8216
27	729	62	3844	97	9409
28	784	63	3969	98	9604
29	841	64	4096	99	9801
30	900	65	4225	100	10000
31	961	66	4356		
32	1024	67	4489		
33	1089	68	4624		
34	1156	69	4761		
35	1225	70	4900		

FIGURE 15.47 ■ Squares of numbers

Diameter	Circumference	Diameter	Circumference
⅛	0.3927	10	31.41
¼	0.7854	10½	32.98
⅜	1.178	11	34.55
½	1.570	11½	36.12
⅝	1.963	12	37.69
¾	2.356	12½	39.27
⅞	2.748	13	40.84
1	3.141	13½	42.41
1⅛	3.534	14	43.98
1¼	3.927	14½	45.55
1⅜	4.319	15	47.12
1½	4.712	15½	48.69
1⅝	5.105	16	50.26
1¾	5.497	16½	51.83
1⅞	5.890	17	53.40
2	6.283	17½	54.97
2¼	7.068	18	56.54
2½	7.854	18½	58.11
2¾	8.639	19	56.69
3	9.424	19½	61.26
3¼	10.21	20	62.83
3½	10.99	20½	64.40
3¾	11.78	21	65.97
4	12.56	21½	67.54
4½	14.13	22	69.11
5	15.70	22½	70.68
5½	17.27	23	72.25
6	18.84	23½	73.82
6½	20.42	24	75.39
7	21.99	24½	76.96
7½	23.56	25	78.54
8	25.13	26	81.68
8½	26.70	27	84.82
9	28.27	28	87.96
9½	29.84	29	91.10
		30	94.24

FIGURE 15.48 ■ Circumference of circle

Diameter	Area	Diameter	Area
⅛	0.0123	10	78.54
¼	0.0491	10½	86.59
⅜	0.1104	11	95.03
½	0.1963	11½	103.86
⅝	0.3068	12	113.09
¾	0.4418	12½	122.71
⅞	0.6013	13	132.73
1	0.7854	13½	143.13
1⅛	0.9940	14	153.93
1¼	1.227	14½	165.13
1⅜	1.484	15	176.71
1½	1.767	15½	188.69
1⅝	2.073	16	201.06
1¾	2.405	16½	213.82
1⅞	2.761	17	226.98
2	3.141	17½	240.52
2¼	3.976	18	254.46
2½	4.908	18½	268.80
2¾	5.939	19	283.52
3	7.068	19½	298.60
3¼	8.295	20	314.16
3½	9.621	20½	330.06
3¾	11.044	21	346.36
4	12.566	21½	363.05
4½	15.904	22	380.13
5	19.635	22½	397.60
5½	23.758	23	415.47
6	28.274	23½	433.73
6½	33.183	24	452.39
7	38.484	24½	471.43
7½	44.178	25	490.87
8	50.265	26	530.93
8½	56.745	27	572.55
9	63.617	28	615.75
9½	70.882	29	660.52
		30	706.86

FIGURE 15.49 ■ Area of circle

Number	Square
1	1
2	4
3	9
4	16
5	25
6	36
7	49
8	64
9	81
10	100
11	121
12	144
13	169
14	196
15	225
16	256
17	289
18	324
19	361
20	400
21	441
22	484
23	529
24	576
25	625
26	676
27	729
28	784
29	841
30	900
31	961
32	1024
33	1089
34	1156
35	1225

Number	Square
36	1296
37	1369
38	1444
39	1521
40	1600
41	1681
42	1764
43	1849
44	1936
45	2025
46	2116
47	2209
48	2304
49	2401
50	2500
51	2601
52	2704
53	2809
54	2916
55	3025
56	3136
57	3249
58	3364
59	3481
60	3600
61	3721
62	3844
63	3969
64	4096
65	4225
66	4356
67	4489
68	4624
69	4761

Number	Square
70	4900
71	5041
72	5184
73	5329
74	5476
75	5625
76	5776
77	5929
78	6084
79	6241
80	6400
81	6561
82	6724
83	6889
84	7056
85	7225
86	7396
87	7569
88	7744
89	7921
90	8100
91	8281
92	8464
93	8649
94	8836
95	9025
96	9216
97	9409
98	9604
99	9801
100	10000

FIGURE 15.50 ■ Squares of numbers

Number	Square root	Number	Square root	Number	Square root
1	1.00000	36	6.00000	71	8.42614
2	1.41421	37	6.08276	72	8.48528
3	1.73205	38	6.16441	73	8.54400
4	2.00000	39	6.24499	74	8.66025
5	2.23606	40	6.32455	75	8.71779
6	2.44948	41	6.40312	76	8.77496
7	2.64575	42	6.48074	77	8.83176
8	2.82842	43	6.55743	78	8.88819
9	3.00000	44	6.63324	79	8.94427
10	3.16227	45	6.70820	80	9.00000
11	3.31662	46	6.78233	81	9.05538
12	3.46410	47	6.85565	82	9.11043
13	3.60555	48	6.92820	83	9.11043
14	3.74165	49	7.00000	84	9.16515
15	3.87298	50	7.07106	85	9.21954
16	4.00000	51	7.14142	86	9.27361
17	4.12310	52	7.21110	87	9.32737
18	4.24264	53	7.28010	88	9.38083
19	4.35889	54	7.34846	89	9.43398
20	4.47213	55	7.41619	90	9.48683
21	4.58257	56	7.48331	91	9.53939
22	4.69041	57	7.54983	92	9.59166
23	4.79583	58	7.61577	93	9.64365
24	4.89897	59	7.68114	94	9.69535
25	5.00000	60	7.74596	95	9.74679
26	5.09901	61	7.81024	96	9.79795
27	5.19615	62	7.87400	97	9.84885
28	5.29150	63	7.93725	98	9.89949
29	5.38516	64	8.00000	99	9.94987
30	5.47722	65	8.06225	100	10.00000
31	5.56776	66	8.12403		
32	5.65685	67	8.18535		
33	5.74456	68	8.24621		
34	5.83095	69	8.30662		
35	5.91607	70	8.36660		

FIGURE 15.51 ▪ Square roots of numbers

Number	Cube	Number	Cube	Number	Cube
1	1	36	46656	71	357911
2	8	37	50653	72	373248
3	27	38	54872	73	389017
4	64	39	59319	74	405224
5	125	40	64000	75	421875
6	216	41	68921	76	438976
7	343	42	74088	77	456533
8	512	43	79507	78	474552
9	729	44	85184	79	493039
10	1000	45	91125	80	512000
11	1331	46	97336	81	531441
12	1728	47	103823	82	551368
13	2197	48	110592	83	571787
14	2477	49	117649	84	592704
15	3375	50	125000	85	614125
16	4096	51	132651	86	636056
17	4913	52	140608	87	658503
18	5832	53	148877	88	681472
19	6859	54	157464	89	704969
20	8000	55	166375	90	729000
21	9621	56	175616	91	753571
22	10648	57	185193	92	778688
23	12167	58	195112	93	804357
24	13824	59	205379	94	830584
25	15625	60	216000	95	857375
26	17576	61	226981	96	884736
27	19683	62	238328	97	912673
28	21952	63	250047	98	941192
29	24389	64	262144	99	970299
30	27000	65	274625	100	1000000
31	29791	66	287496		
32	32768	67	300763		
33	35937	68	314432		
34	39304	69	328500		
35	42875	70	343000		

FIGURE 15.52 ■ Cubes of numbers

Diameter	Area	Diameter	Area
⅛	0.0123	10½	86.59
¼	0.0491	11	95.03
⅜	0.1104	11½	103.86
½	0.1963	12	113.09
⅝	0.3068	12½	122.71
¾	0.4418	13	132.73
⅞	0.6013	13½	143.13
1	0.7854	14	153.93
1⅛	0.9940	14½	165.13
1¼	1.227	15	176.71
1⅜	1.484	15½	188.69
1½	1.767	16	201.06
1⅝	2.073	16½	213.82
1¾	2.405	17	226.98
1⅞	2.761	17½	240.52
2	3.141	18	254.46
2¼	3.976	18½	268.80
2½	4.908	19	283.52
2¾	5.939	19½	298.6
3	7.068	20	314.16
3¼	8.295	20½	330.06
3½	9.621	21	346.36
3¾	11.044	21½	363.05
4	12.566	22	380.13
4½	15.904	22½	397.60
5	19.635	23	415.47
5½	23.758	23½	433.73
6	28.274	24	452.39
6½	33.183	24½	471.43
7	38.484	25	490.87
7½	44.178	26	530.93
8	50.265	27	572.55
8½	56.745	28	615.75
9	63.617	29	660.52
9½	70.882	30	706.89

FIGURE 15.53 ■ Area of a circle

Diameter	Circumference	Diameter	Circumference
⅛	.3927	10½	32.98
¼	.7854	11	34.55
⅜	1.178	11½	36.12
½	1.570	12	37.69
⅝	1.963	12½	39.27
¾	2.356	13	40.84
⅞	3.748	13½	42.41
1	3.141	14	43.98
1⅛	3.534	14½	45.55
1¼	3.927	15	47.12
1⅜	4.319	15½	48.69
1½	4.712	16	50.26
1⅝	5.105	16½	51.83
1¾	5.497	17	53.40
1⅞	5.890	17½	54.97
2	6.283	18	56.54
2¼	7.068	18½	58.11
2½	7.854	19	59.69
2¾	8.639	19½	61.26
3	9.424	20	62.83
3¼	10.21	20½	64.40
3½	10.99	21	65.97
3¾	11.78	21½	67.54
4	12.56	22	69.11
4½	14.13	22½	70.68
5	15.70	23	72.25
5½	17.27	23½	73.82
6	18.84	24	75.39
6½	20.42	24½	76.96
7	21.99	25	78.54
7½	23.56	26	81.68
8	25.13	27	84.82
8½	26.70	28	87.96
9	28.27	29	91.10
9½	29.84	30	94.24
10	31.41		

FIGURE 15.54 ■ Circumference of a circle

Decimal equivalent	Millimeters
.0625	1.59
.1250	3.18
.1875	4.76
.2500	6.35
.3125	7.94
.3750	9.52
.4375	11.11
.5000	12.70
.5625	14.29
.6250	15.87
.6875	17.46
.7500	19.05
.8125	20.64
.8750	22.22
.9375	23.81
1.000	25.40

FIGURE 15.55 ■ Decimals to millimeters

Inches	Decimal of a foot	Inches	Decimal of a foot
⅛	.01042	1⅞	.15625
¼	.02083	2	.16666
⅜	.03125	2⅛	.17708
½	.04167	2¼	.18750
⅝	.05208	2⅜	.19792
¾	.06250	2½	.20833
⅞	.07291	2⅝	.21875
1	08333	2¾	.22917
1⅛	.09375	2⅞	.23959
1¼	.10417	3	.25000
1⅜	.11458		
1½	.12500		
1⅝	.13542		
1¾	.14583		

Note: To change inches to decimals of a foot, divide by 12. To change decimals of a foot to inches, multiply by 12.

FIGURE 15.56 ■ Inches converted to decimals of feet

Fractions	Decimal equivalent
¹⁄₁₆	.0625
⅛	.1250
³⁄₁₆	.1875
¼	.2500
⁵⁄₁₆	.3125
⅜	.3750
⁷⁄₁₆	.4375
½	.5000
⁹⁄₁₆	.5625
⅝	.6250
¹¹⁄₁₆	.6875
¾	.7500
¹³⁄₁₆	.8125
⅞	.8750
¹⁵⁄₁₆	.9375
1	1.000

FIGURE 15.57 ■ Fractions to decimals

Fraction	Decimal		Fraction	Decimal
1/64	.015625		13/32	.40625
1/32	.03125		27/64	.421875
3/64	.046875		7/16	.4375
1/20	.05		29/64	.453125
1/16	.0625		15/32	.46875
1/13	.0769		31/64	.484375
5/64	.078125		1/2	.5
1/12	.0833		33/64	.515625
1/11	.0909		17/32	.53125
3/32	.09375		35/64	.546875
1/10	.10		9/16	.5625
7/64	.109375		37/64	.578125
1/9	.111		19/32	.59375
1/8	.125		39/64	.609375
9/64	.140625		5/8	.625
1/7	.1429		41/64	.640625
5/32	.15625		21/32	.65625
1/6	.1667		43/64	.671875
11/64	.171875		11/16	.6875
3/16	.1875		45/64	.703125
1/5	.2			
13/64	.203125			
7/32	.21875			
15/64	.234375			
1/4	.25			
17/64	.265625			
9/32	.28125			
19/64	.296875			
5/16	.3125			
21/64	.328125			
1/3	.333			
11/32	.34375			
23/64	.359375			
3/8	.375			
25/64	.390625			

FIGURE 15.58 ■ Decimal equivalents of fractions

Minutes	Decimal of a degree
1	.0166
2	.0333
3	.0500
4	.0666
5	.0833
6	.1000
7	.1166
8	.1333
9	.1500
10	.1666
11	.1833
12	.2000
13	.2166
14	.2333
15	.2500
16	.2666
17	.2833
18	.3000
19	.3166
20	.3333
21	.3500
22	.3666
23	.3833
24	.4000
25	.4166

FIGURE 15.59 ■ Minutes converted to decimal of a degree

Fraction	Decimal
1/32	.03125
1/16	.0625
3/32	.09375
1/8	.125
5/32	.15625
3/16	.1875
7/32	.21875
1/4	.25
9/32	.28125
5/16	.3125
11/32	.34375
3/8	.375
13/32	.40625
7/16	.4375
15/32	.46875
1/2	.5
17/32	.53125
9/16	.5625
19/32	.59375
5/8	.625
21/32	.65625
11/16	.6875
23/32	.71875
3/4	.75
25/32	.78125
13/16	.8125
27/32	.84375
7/8	.875
29/32	.90625
15/16	.9375
31/32	.96875
1	1.000

FIGURE 15.60 ■ Decimal equivalents of an inch

Inches	Decimal of an inch
¹⁄₆₄	.015625
¹⁄₃₂	.03125
³⁄₆₄	.046875
¹⁄₁₆	.0625
⁵⁄₆₄	.078125
³⁄₃₂	.09375
⁷⁄₆₄	.109375
¹⁄₈	.125
⁹⁄₆₄	.140625
⁵⁄₃₂	.15625
¹¹⁄₆₄	.171875
³⁄₁₆	.1875
¹²⁄₆₄	.203125
⁷⁄₃₂	.21875
¹⁵⁄₆₄	.234375
¹⁄₄	.25
¹⁷⁄₆₄	.265625
⁹⁄₃₂	.28125
¹⁹⁄₆₄	.296875
⁵⁄₁₆	.3125

Note: To find the decimal equivalent of a fraction, divide the numerator by the denominator.

FIGURE 15.61 ■ Decimal equivalents of fractions of an inch

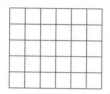

INDEX

Page numbers in italics refer to figures and tables